Forward

The establishment of the National Network for Pulp and Paper Technology Training, (npt)^2, marks the beginning of a new era of workforce training and education for the U.S. pulp and paper industry. As the industry continues its quest to advance innovative technologies to strengthen its global competitive position, the urgency of the need for significant upgrading of the level of technical proficiency of the operating and maintenance workforce intensifies. (npt)^2 is built upon a new model of human resource development in the U.S. involving strong partnerships among the operating companies and their mills, industry associations, universities and community/technical colleges. The network, with generous support from the National Science Foundation and a growing number of companies, now has operating curricula, in the form of two-year associate degree programs, in place in community colleges in all the major pulp and paper manufacturing regions of the nation. With advice and assistance from a number of universities of PPERA (The Pulp and Paper Education and Research Alliance), these programs have been created using the vision of Agenda 2020's Technologically Advanced Workforce (TAW) platform as a guiding principle. TAPPI, with its long history as the leading provider of technical education for industry professionals, has provided access to its vast reservoir of expertise to assist in curricular development and information dissemination.

The role of TAPPI in (npt)^2 takes on particular importance as the need for support for the programs in the form of quality textbooks is addressed. (npt)^2 partners, with assistance from PPERA university faculty, were able to leverage trusted sources of peer-reviewed material from the TAPPI knowledge base. Beginning with the TAPPI eLearning series *Making Pulp and Paper*, and drawing also from learning material available through PPERA universities, (npt)^2 community colleges, and TAPPI short courses, the authors created Volume 1 "*Pulp Manufacturing,*" of the *Pulp and Paper Technology Advanced Workforce Training and Education Series* textbooks.

The *Pulp and Paper Technology Advanced Workforce Training and Education* textbook series will be made up of three volumes to be published in consecutive years. Volume 1 contains all major areas of pulping and bleaching, including wood preparation, chemical recovery and fiber recycling. Following publication of Volume 1 in September 2006 will be Volume 2 - *Paper Manufacturing*, scheduled for publication in July 2007. Volume 3 is scheduled for publication in 2008 and is expected to cover safety, the environment, and other topics needed to support an advanced pulp and paper workforce.

While the textbook series is designed primarily to be an important resource for community college students, manufacturing and workforce development leaders will find it to be an invaluable tool to improve new and incumbent worker knowledge.

The *Pulp and Paper Technology Advanced Workforce Training and Education* series is an important project funded through a major NSF-ATE grant supporting the formation of (npt)^2. These textbooks are uniquely designed to support the basic pulp and paper technology curriculum model adopted by the member colleges of (npt)^2. As additional colleges are added to the network, these texts will greatly facilitate the deployment of many more programs of similar structure and content to meet our industry's goal of developing a technologically advanced workforce.

— *Dr. Harry T. Cullinan, Jr.*
Director, Alabama Center for Paper and Bioresource Engineering
Auburn University

Acknowledgements

Production of this book is supported by The National Science Foundation, Advanced Technological Educational Center of Excellence program through The National Network for Pulp and Paper Technology Training.

The Pulp and Paper Technology Advanced Workforce Training and Education Series

TAPPI PRESS
15 Technology Parkway South
Norcross, GA 30092
United States

TAPPI PRESS
15 Technology Parkway South
Norcross, GA 30092 U.S.A.
www.tappi.org

ISBN: 1-59510-133-0
TAPPI PRESS Order Number: 0101R317
Printed in the United States of America

Affiliations

The Pulp and Paper Technology Advanced Workforce Training and Education Series
Volume 1: 2006

Editor-in-Chief
Dr. Harry Cullinan, Jr., Auburn University

Chapter 1: Wood Fibers
Karyn Biasca, University of Wisconsin - Stevens Point

Chapter 2: Pulp Mill Overview
Laura Feix, Consultant and Harry Cullinan, Auburn University

Chapter 3: Wood Preparation
Ralph Benefiel, Lower Columbia College

Chapter 4: Kraft Pulping
Jim Bond, Mid-State Technical College and Karyn Biasca, University of Wisconsin - Stevens Point

Chapter 5: Chemical Recovery and Energy Production
Dale Smith, Alabama Southern Community College and Gopal Krishnagopalan, Auburn University

Chapter 6: Mechanical and Chemi-Mechanical Pulping
Monty Henderson, Kennebec Valley Technical College

Chapter 7: Bleaching
Bob Caron, TAPPI

Chapter 8: Recycling and Deinking
Steve Duke and Gopal Krishnagopalan, Auburn University

Cover and Layout Design
Magdalena Baboolal and Deborah Robinson

Cover Art Images
Courtesy of Stora Enso

Image Contributor
Metso Paper

Table of contents

Chapter 1

WOOD FIBERS

OBJECTIVES

After completing this section, the student will be able to identify and explain:

- ***Why wood fibers make good papermaking materials.***
- ***The functions of different types of cells in the tree.***
- ***What wood fibers look like.***
- ***Some of the differences between hardwoods and softwoods.***

1.1 KEY TERMS

Bagasse - the crushed stalks of sugar cane, which are pulped and made into paper.

Chlorophyll - the green chemical found in plants that is necessary for photosynthesis to occur.

Deciduous trees - hardwood or broadleaf trees that lose their leaves annually.

Glucose - a simple sugar produced by photosynthesis in green plants.

Hardwoods - trees such as birch, maple, oak, eucalyptus, and gum, characterized by broad leaves and short, slender fibers.

Lignin - a substance that acts as a glue between cellulose molecules to form the woody cell wall of plants.

Lumen - the void found in the middle of plant fibers.

Photosynthesis - the process by which green plants convert carbon dioxide and water into glucose in the presence of sunlight.

Softwoods - wood from coniferous trees characterized by needle-like leaves, such as pine, spruce, or hemlock. The fibers in softwoods are longer than hardwood fibers and make stronger paper.

Tracheids - the long, tubular cells in wood that collect moisture and support the stem. Commonly called fibers in the papermaking process.

Vessels - a type of large cell found in hardwoods which does not bond well in the papermaking process.

1.2 INTRODUCTION

When you tear a sheet of paper, the torn edge has a fuzzy appearance, as shown in Figure 1. Upon closer inspection, you will see that the sheet of paper is actually made up of tiny fibers. In most grades of paper, these fibers come from wood. Wood fibers are especially well suited for papermaking because of their distinctive properties. In this chapter, we will explore the structure of the trees that give us these fibers, what the fibers look like, how they function in the tree, and why they make good paper.

Figure 1: Notice the fibers on the torn edge of a sheet of paper

Though fibers have many important properties, there are two that are of primary importance for papermaking. The fibers must bond to each other on their own, without the addition of glues. The lack of this ability makes most synthetic fibers difficult to use for making paper. The fibers must also be able to form the thin, layered structure that we recognize as paper. All plant fibers have both of these properties. Non-wood fibers, such as cotton, flax and bagasse, are suitable for some types of paper. A major disadvantage of most non-wood fibers is that they are annual crops, resulting in material handling issues associated with their use. Non-wood plants may also contain large amounts of extraneous materials that must be removed before the fibers are suitable for papermaking. Fibers from wood have significant advantages and provide most of the fiber used in paper mills today. We will begin our study of wood fibers by looking at the structure of trees.

1.3 TREES

Trees, in addition to creating a beautiful environment in the forests where they grow, contain a remarkable "chemical factory". Using energy from the sun combined with the green chemical chlorophyll, trees absorb carbon dioxide from the air and combine it with water that comes from the ground to produce glucose in the photosynthesis process (Figure 2). Glucose is a simple sugar that is the building block for almost everything in the tree. The tree makes cellulose, the structural backbone of the fibers, from glucose. Hemicelluloses, a "filler" material, are also produced from glucose and other simple sugars in the tree. Lignin, the glue that holds the entire structure together, is also produced through complicated enzymatic pathways that use glucose as the starting material. These chemicals are interwoven to produce the structures within the trees that keep them upright, move water from the ground to the leaves, and perform other functions important to the tree's well-being.

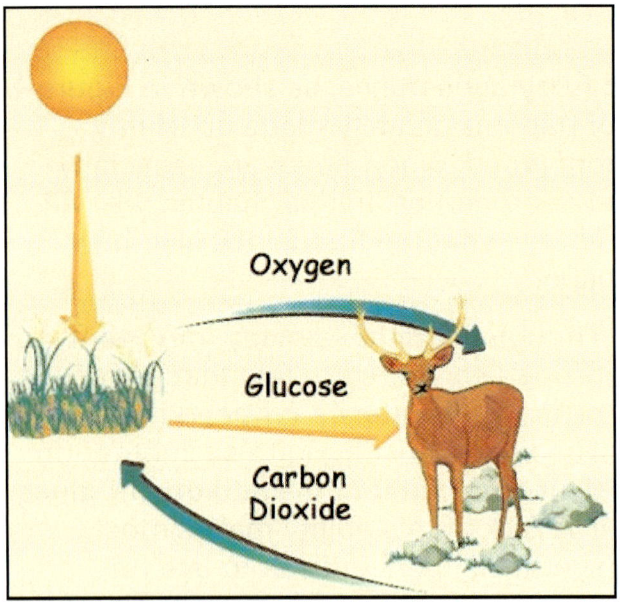

Figure 2: Ecosystems depend on Photosynthesis

There are thousands of different species of trees in the world, and each species differs from the others in significant ways. For our purposes, however, we will groups trees into two major categories: softwoods and hardwoods. From a papermaking point of view, the differences between softwoods and hardwoods are much more significant than the differences among softwood species or among hardwood species. In general, softwood species hold on to their (usually) needle-shaped leaves throughout the winter; the seeds for these trees are often borne in cones. For these reasons, softwoods are sometimes referred to as "evergreens" or "conifers". Hardwoods, on the other hand, typically drop their leaves in the fall. The leaves are usually broad and flat. These characteristics lead to the common names of "deciduous" or "broadleaf" for hardwoods. From a papermaker's point of view, the major difference between softwoods and hardwoods is the size and shape of the fibers used for paper (Figure 3). Softwoods have long (3-5 mm), strong fibers that lend themselves to the production of high-strength papers. These papers are generally rough and not well suited for high-quality printing (Figure 4a). Hardwoods, on the other hand, have short (about 1 mm), flexible fibers that produce a very smooth surface on a sheet of paper (Figure 4b). Using mixtures of these two types of fibers in different proportions, the papermaker can craft a wide variety of papers, suitable for many purposes.

Figure 3: Comparison of softwood and hardwood fibers

1.4 FIBERS

Let's look a bit more closely at the structure of these fibers (Figure 5). The fibers are hollow, tubular structures with closed, tapered ends. They look a bit like drinking straws that have been closed and flattened on the ends. The hollow area inside the fiber is called the "lumen". The wall thickness of the fibers may vary greatly. The wall thickness is an important contributor to the strength of the fibers in the wood and their ability to keep the tree upright and growing toward the sunlight.

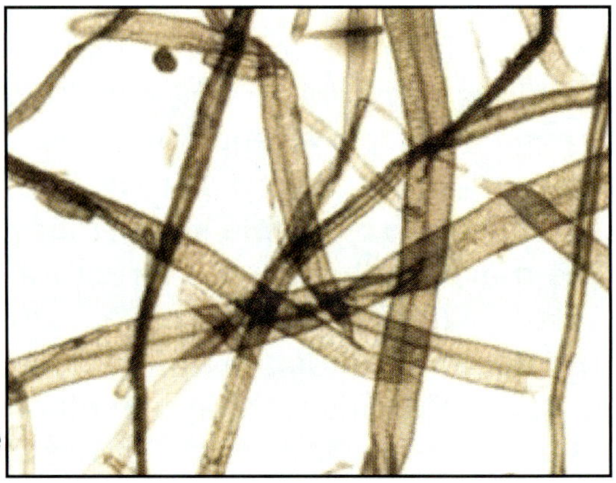

Figure 5

Figure 4a: Softwood fibers

Figure 4b: Hardwood fibers

In softwoods, the fibers perform another important function for tree growth. There are many small holes in their fiber walls (Figure 6a). These holes are called "pits", and when the fibers are packed together in the wood, the holes in one fiber line up with the holes in the adjacent fibers. This forms a network of connected tubes to allow water to move from the ground up through the stem of the tree to the leaves (needles), where it is needed for photosynthesis.

The conduction of water from the roots to the leaves is accomplished by a different set of structures in hardwood trees. Hardwoods have vessel cells (Figure 6b), which are much shorter and wider than the fibers. These vessel cells are stacked one on top of another to form a tube-like structure that provides a path for water conduction. These vessel cells can

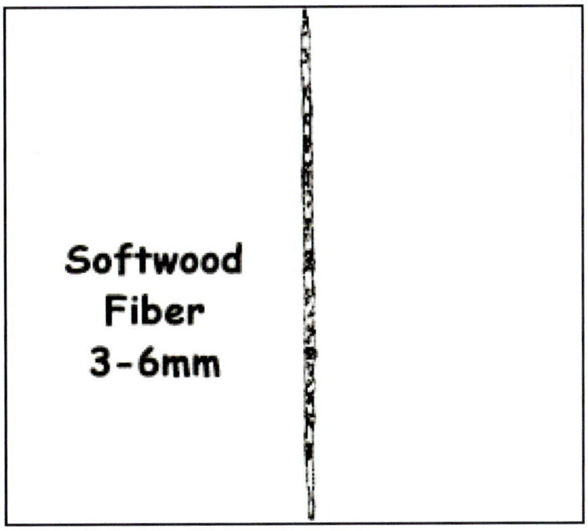

Figure 6a: Software fiber

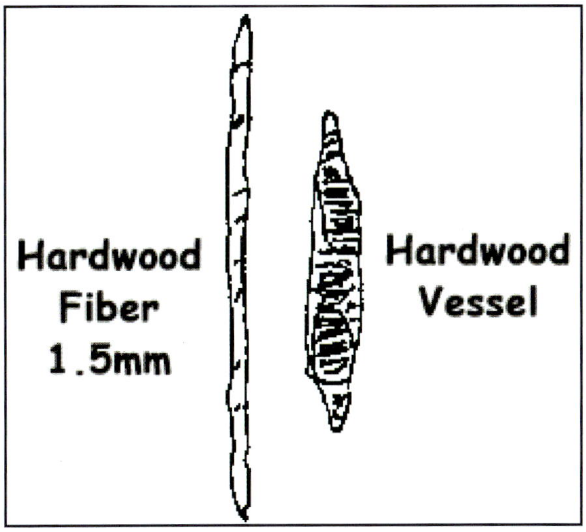

Figure 6b: Hardwood fiber

be problematic in papermaking, since they do not bond in the sheet as well as fibers do. The vessels have a tendency to pick out of the sheet whenever they come into contact with a sticky surface.

Although the fibers are quite stiff when inside the wood, they have a layered structure (Figure 7) which allows them to be mechanically treated to make them flexible and pliable, resulting in good bonding characteristics for making paper. The highest sheet strength is achieved when the fibers are pliable enough to collapse into thin, ribbon-like structures, giving the maximum area of contact for bonding (Figure 8).

It should be noted that there are major differences in fibers between different species of softwoods or different species of hardwoods. Generally, because of the cost of transporting wood to a mill, the most abundant species of trees near the mill are the ones used for paper production. Each species has its own properties, and these properties are reflected in the paper made from the trees. So the situation is a lot more complex than just softwood or hardwood.

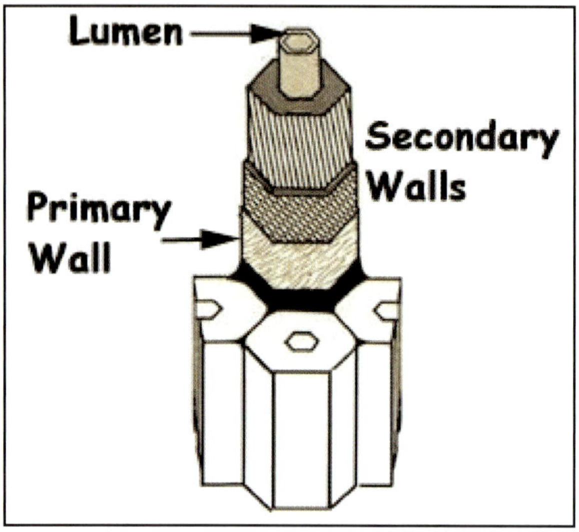

Figure 7

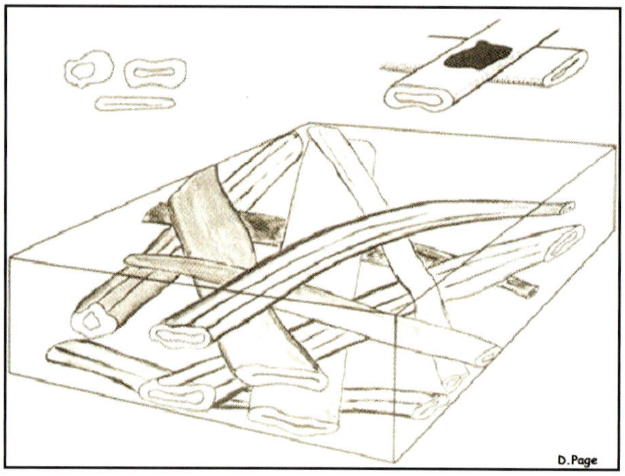

Figure 8

1.5 REVIEW QUESTIONS

You will know that you have understood the materials in this chapter if you can easily answer the following questions:

1. What is photosynthesis?
   A. A process by which green plants turn carbon dioxide and water into sugar.
   B. An early type of photographic process.
   C. A way to store photographic images digitally.
   D. A chemical reaction in which chlorophyll and oxygen are converted into glucose.

2. What are some of the differences between hardwood and softwood fibers?
   A. Hardwood fibers are much longer than softwood fibers.
   B. Softwood fibers do not bond as well and form weaker paper.
   C. Softwood fibers have a lumen and hardwood fibers do not.
   D. Hardwood fibers are much shorter and are better for paper printability.

3. What is lignin?
   A. A unit of measurement for wood.
   B. The glue that hold fibers together in the tree.
   C. A material that occurs in softwood but not in hardwoods.
   D. The hole in the middle of the fiber.

Chapter 2

PULP MILL OVERVIEW

OBJECTIVES

At the conclusion of this section the student will be able to:

- ***Explain the major types of pulping processes.***
- ***List one advantage and one disadvantage of chemical pulping as compared to mechanical pulping.***
- ***Describe in general terms the processing of pulp after removal from the digester.***
- ***Describe in general terms the processing of the spent cooking liquor.***

2.1 KEY TERMS

Black Liquor - The spent cooking liquid from the kraft pulping process.

Brownstock - The pulp slurry after washing but before any bleaching process. It is brown because of the remaining lignin in the fibers.

Cellulose - The main structural constituent of woody plants. Wood cellulose is the material remaining after much of the lignin and certain carbohydrates have been removed from wood cell walls by pulping and bleaching.

Chemical Pulping - A process to dissolve lignin by the use of chemicals in order to liberate the cellulose fibers.

Chlorine - Cl_2. A bleaching chemical that removes the colored lignin in brownstock.

Clarifier - A type of tank used in chemical recovery to clarify green or white liquor by allowing entrained particles to settle out of the main upward flow and be removed as slurry from the bottom of the tank.

Concentrator - An evaporator supplied with fresh steam, used in the final stage of concentration of an aqueous solution.

Consistency - The percentage by weight of ovendry fibrous material in a stock or stock suspension. It is sometimes called density, solids, or concentration.

Cooking Liquor - The solution of chemicals used in pulping to remove the lignin. Depending upon the pulping stage, the liquor may be white (fresh kraft cooking liquor), black (after pulping), or green (after chemical recovery but before causticizing).

Digester - A vessel that is used to pulp fibrous raw materials such as wood chips by chemical means.

Dioxin - The common term for a family of chlorinated organic compounds that are mildly to highly toxic. Can be produced in the bleaching process if chlorine is present.

Disc Refiners - A piece of rotating equipment with two or more discs which have bars on their surfaces. The discs are set close together so that wood chips that pass between the two surfaces are separated into fibers.

Evaporator - Equipment to remove water from an aqueous solution (e.g. black liquor) to produce a solution of higher concentration.

Groundwood - a high-yield mechanical pulp produced by pressing logs sideways against a rotating pulp stone.

Hemicellulose - a broad class of polysaccharides which are associated with cellulose in plant cell walls. Hemicelluloses have a lower molecular weight than cellulose.

Kraft Pulping - an alkaline pulping process that uses sodium hydroxide and sodium sulfide to dissolve lignin and release cellulose fibers.

Lignin - a substance that acts as a glue between cellulose fibers to form the woody structure of plants.

Lime Mud - Slurry or cake of fine particles of calcium carbonate derived from the reaction of reburned or fresh lime with green liquor.

Mechanical Pulping - A high-yield pulping process which depends primarily on mechanical action to separate fibers with minimal removal of lignin.

Pulping -The process of separating a cellulose-containing raw material, such as wood or straw, into individual fibers which can be reformed to make paper.

Recausticizing - A part of the kraft recovery process in which lime is added to green liquor to form white liquor.

Slaker - An agitated tank in which lime is reacted with green liquor or water to form calcium hydroxide.

2.2 INTRODUCTION TO PULPING METHODS

Fibers in wood or other cellulose-containing materials are held together by a chemical glue called lignin. In North America, the main source of fiber is wood, but in other parts of the world, sources of fiber include reeds (bagasse), bamboo, or even grasses. In the United States and Canada, where wood is used, pulping either dissolves the lignin with chemicals to liberate the fibers (chemical pulping), separates the wood fibers physically (mechanical pulping), or involves a combination of chemical and mechanical pulping (Semi-chemical pulping). Once you have the individual pulp fibers, you can make paper or paperboard. Different pulping processes produce different kinds of pulp which are used in a wide range of products. For example, mechanical pulps are used in the manufacture of newsprint, phone books, and "throw-away" catalogues. Chemical pulps are most commonly utilized in books of all kinds, high-quality magazines, annual reports, and the like. Semi-chemical pulps are ideally suited for corrugating medium - the fluted material in cardboard that gives boxes their strength. Cellulose itself, the main chemical component of pulp fibers, can be made into products such as cellulose acetate film or rayon textile fiber. The range of products manufactured is huge, but they all start with some type of pulping process.

2.2.1 Mechanical Pulping

Cellulose fibers from plants such as cotton or linen have been used to make paper for thousands of years. Only in the middle of the nineteenth century did we begin to make pulp from wood. One of the first processes was groundwood pulping, which is still in use today. "Groundwood" is both a name and a literal description. Logs are pressed against a wet, rough, rotating grindstone. The friction heats the wood and softens the lignin, and fibers or pieces of fibers are broken off. Groundwood pulping is the classical mechanical pulping process, but it has been superseded in the past three decades by mechanical processes which use disc to convert wood chips into pulp. You'll learn more about them in Chapter 6.

As you might expect, grinding or refining can easily damage the fibers, but little or no lignin is actually removed. The fibers are short and comparatively weak, but they are suitable, when mixed with longer fibers, for the production of paper for newspapers, advertising flyers, and magazines. The great thing about groundwood and all of the mechanical pulping processes is that the yield of pulp from wood is very high, typically 90% or better. In other words, 90% of the wood you put into the pulping process comes out as pulp.

2.2.2 Semi-chemical Pulping

Mechanical pulps are fine for many grades of paper, but their relatively low strength can be a disadvantage. One way to improve the quality of the pulp is to start with a short chemical cook and finish with a mechanical pulping process. In these hybrid processes, some of the lignin is first removed chemically, so that the fibers can then be separated with less mechanical action. This results in lower yield, but the pulp fibers are stronger. A good example is the production of pulp for corrugating medium, the fluted material between the two layers of linerboard forming the walls of a corrugated box. These hybrid pulping processes are classified as Semi-chemical or Chemi-mechanical. The most common Semi-chemical process is called Neutral Sulfite Semi-chemical (NSSC). Wood can also be partially cooked using the kraft or sulfite chemical processes (described below).

2.2.3 Chemical Pulping

Both pure mechanical pulping processes and hybrid ones leave most of the lignin in the pulp. This impairs the strength of paper made from such pulps and causes yellowing with exposure to light and air. To produce really strong pulp fibers which can also be bleached to high and stable whiteness, we need to remove almost all of the lignin. That's what happens in chemical pulping processes, which separate wood fibers by dissolving the lignin that holds them together. Wood chips are heated in a chemical solution inside a pressure vessel called a digester. At high temperature, the chemicals dissolve the lignin, eventually liberating the fibers. At the end of the cook, the material is discharged from the digester and emerges as pulp.

The two main chemical pulping processes in use today are sulfite and kraft. Kraft pulping is by far the dominant one, accounting for more than 70% of all pulp and 90% of all chemical pulp produced worldwide. The two processes are chemically different. Sulfite pulping, actually a family of processes, uses bisulfite solutions and varying amounts of sulfurous acid under acidic conditions to dissolve the lignin. Kraft pulping, at the other end of the pH scale, uses a mixture of sodium hydroxide and sodium sulfide under strongly alkaline conditions. Both processes have been around for a long time - kraft is about 120 years old, sulfite even older.

The kraft process is always coupled to a chemical recovery system. Chemical recovery continuously reprocesses dissolved inorganic material in the spent pulping liquor into fresh cooking liquor for pulping. It also generates a considerable amount of energy for steam and power production from the dissolved lignin. Without chemical recovery, kraft pulping would be uneconomical - and an environmental nightmare! Nowadays, the various sulfite pulping processes also have complete chemical recovery systems, but that wasn't always the case in the past.

Chemical pulping produces pulps whose fibers bond well to one another to form strong paper. However, the pulp yield is much lower than in mechanical or hybrid pulping processes. While yields in many mechanical pulping processes are 90% or better, those in chemical processes are typically 45% to 55%, so it takes a lot more wood to produce pulp this way. Most of the material that is removed in chemical pulping is lignin and extractives, but inevitably there is also some loss of hemicelluloses and cellulose, the other two main components of wood.

Chemical pulping is done in huge "pressure cookers" called digesters. The process is rather like making beef stew - you cut up the solid ingredients (beef, carrots, potatoes, and so forth) into pieces, add a certain amount of water according to the recipe, put it all in a pressure cooker, and heat it at a given temperature for a certain amount of time. And out comes your stew - or kraft pulp, in this case.

There are two kinds of digesters for kraft pulping - batch digesters and continuous digesters. Continuous digesters are much more common because they are more energy-efficient than batch digesters.

In a batch digester, illustrated in Figure 1, chips are fed into the top of the digester and cooking liquor is added. Once the digester is sealed, the digester is brought up to the desired temperature and the chips are allowed to cook for the prescribed length of time. At the end of batch cooking, the pressure is released through a blow valve at the bottom of the digester, and the cooked chips and spent liquor are discharged into a blow tank. The turbulent discharge flow disintegrates the cooked chips into fibrous pulp.

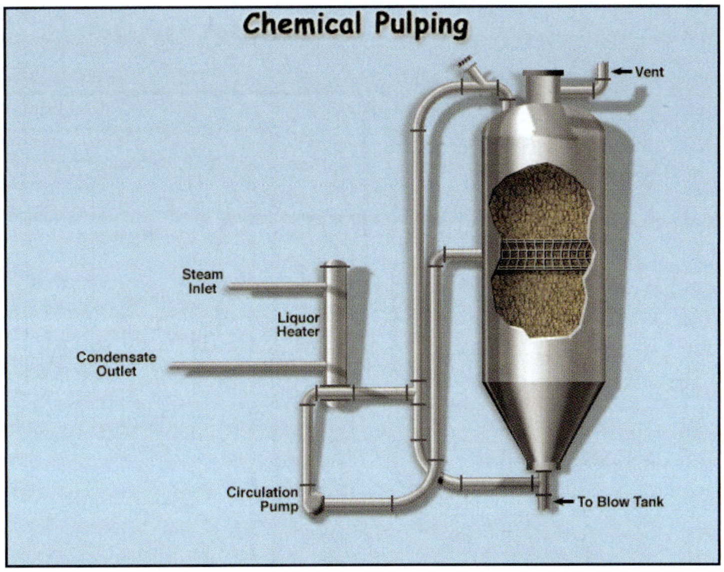

Figure 1: Batch digester

A continuous digester, shown in Figure 2, produces pulp in a continuous stream. The wood chips are pre-steamed, then mixed with cooking liquor and transferred into the pressure vessel. Cooking occurs as the column of chips and liquor is heated to high temperature and slowly descends through the digester. When pulping is complete, hot spent liquor is removed through screens. It is replaced by cooler "wash liquor" which is pumped upward from the bottom of the digester. The cooked material exits from the bottom of the digester through a special outlet device and is discharged into a blow tank or directly to a pulp washer. Figure 3 provides a diagram of the continuous digester system.

Figure 2: Continuous digester

The spent cooking chemicals still with the pulp need to be separated for recovery and reuse. For this reason, the pulp is washed and the residual liquor is removed as wash filtrate. The residual liquor is sent to the chemical recovery side of the mill. There it is concentrated and burned as a fuel and the inorganic cooking chemicals are recovered. Figure 4 shows a pulp washer.

Figure 3: Continuous digester system for kraft cooking

2.3 PULP PROCESSING

After washing, the unbleached pulp is usually screened to remove any material other than individual pulp fibers - uncooked pieces of wood, for example. Pulp screens use perforated barriers to allow acceptable fibers to pass while rejecting unwanted material. Then the pulp is thickened to about 10% consistency (10% solids) and sent to an intermediate storage tank.

Figure 4: Vacuum drum washer

If the final paper product is unbleached - in other words, brown-colored like linerboard or grocery bags - the pulp is diluted and pumped from high-density storage to the paper mill. But many paper products - food board and photocopy paper are examples - must be bright white, so the pulp has to be bleached from brown to white. Although bleachable-grade chemical pulping typically removes more than 90% of the original lignin, the small amount of remaining lignin is responsible for the brown color of unbleached pulp. Not surprisingly, this pulp is usually called brownstock.

Why not simply cook until all of the lignin is gone to get the pulp white and bright? The answer is that, by the time the pulp becomes white, the cellulose is degraded to the point where the pulp is not usable as fiber to make paper products. The fiber has lost most of its strength. So the wood chips are cooked only to a point where there is minimal damage to the cellulose fibers, and the remaining dark-colored lignin is removed later by oxygen delignification and bleaching.

One stage in pulp manufacture has become much more common during the 1980s and 1990s: oxygen delignification, or "O_2 Delig" for short. Oxygen delignification is sometimes referred to as bleaching, although it's really part of the "brown" part of a pulp mill. This is because oxygen delignification primarily removes lignin, rather than brightening the pulp.

Originally, cloth rags were the raw material for paper, and bleaching the rags meant laying them in the sun to fade and then boiling them in potash solutions. This took a long time. But way back in 1774, Carl Scheele discovered chlorine, which could be used to make hypochlorite, a powerful bleaching chemical. You probably have some hypochlorite in your laundry room at home. Liquid laundry bleach, for example, is a 6% solution of hypochlorite. For many decades, chlorine and hypochlorite solutions were the most commonly used bleaching chemicals.

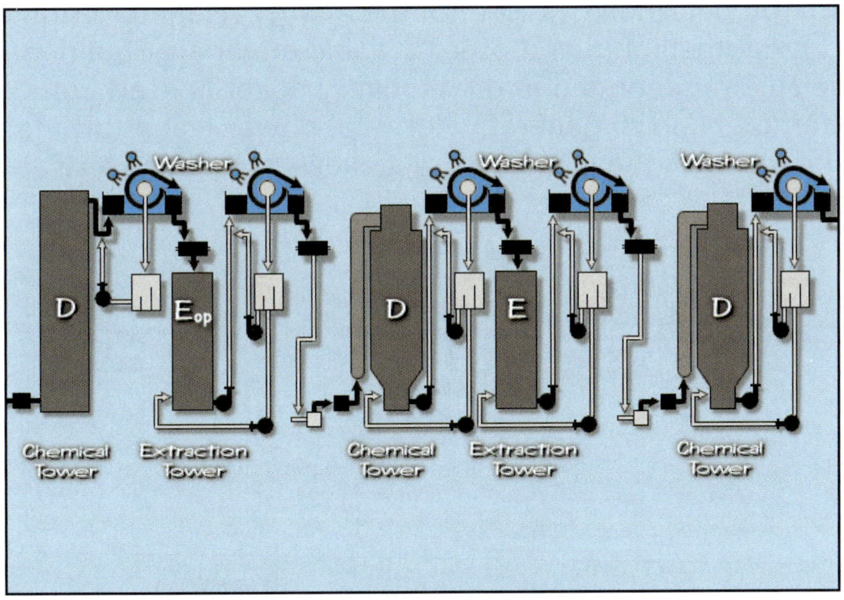

Figure 5: Flow diagram of bleaching process

Pulp bleaching is done in a continuous sequence of stages, as seen in Figure 5. Each stage uses a different chemistry and is usually separated from the next stage by washing. During the late 1980s and early 1990s, scientists discovered that under certain bleaching conditions using chlorine (Cl_2), a class of highly toxic chemicals could be formed in tiny amounts as a result of unwanted side reactions. These compounds were collectively referred to as "dioxins" and "furans" and were highly toxic to laboratory animals. The discovery of these compounds triggered an unprecedented amount of research and development to discover new ways to bleach pulp that would not produce these toxic compounds. Fortunately, scientists found that by eliminating the use of Cl_2 and ensuring that dioxin- and furan-free defoamers were used, the formation of chlorinated dioxins and furans could be virtually eliminated. This is how the term "Elemental Chlorine-Free", or ECF, bleaching came into use.

Some mills eliminated all chlorine-containing bleaching chemicals to achieve totally chlorine-free bleaching, or TCF. Greater use of oxygen, hydrogen peroxide, ozone, and enzymes has expanded the choice of bleaching reagents, but the favorite one by far is chlorine dioxide, a very selective chemical for removing residual lignin and making pulp bright white. After the last bleaching stage, the bleached pulp is thickened and placed in large, high-density storage tanks to be used on the mill's paper machines, or to be run over lap machines and sold to other mills that make paper products.

2.4 CHEMICAL RECOVERY

While the fibers isolated in the digester are processed into a raw material suitable to make paper and board, the liquor that has been separated from the fibers is sent to another process to recover valuable chemicals. This is necessary not only for economic reasons, but for environmental reasons as well. A large part (half or more) of the energy needed to run a pulp and paper mill is obtained by burning the lignin and other organic materials obtained from cooking in a recovery boiler. The steam generated in the recovery boiler is used both as process steam (to dry paper, for example) and to generate electrical power using turbine generators. The flow diagram in Figure 6 shows the steps that take place in the recovery process.

In brownstock washing, black liquor is removed from the pulp and collected. Since it has been diluted with wash water, the weak black liquor is sent to the evaporators and then to a concentrator, where most of the water is removed. The resulting strong black liquor goes to the recovery boiler, where its organic content is burned as a fuel to generate high-pressure steam. Simultaneously, the inorganic chemicals from the black liquor, which fall to the bottom of the boiler, are recovered from the boiler as a molten stream rather like lava flowing from a volcano. The steam generated in the recovery boiler is used to generate electricity in a turbo-generator and thereby reduced to medium- and low-pressure steam for heating the pulping and bleaching operations.

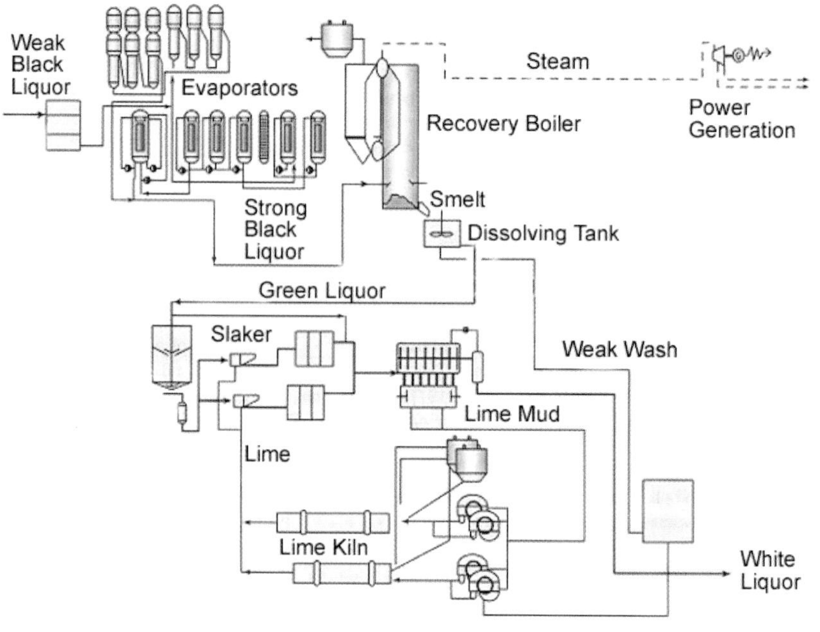

Figure 6: Chemical Recovery Flow Diagram

The molten chemicals, called smelt, are dissolved in weak wash to form green liquor. The green liquor is clarified and sent to the slaker, where it is reacted with hydrated lime to recausticize it to fresh white liquor. This in turn is clarified to produce clean white liquor for use in the digester. The lime mud is washed and sent to the lime kiln for calcining, which converts the lime mud back into lime. The filtrate from mud washing - called weak wash - goes to the dissolving tank which receives the smelt from the recovery boiler. Kraft recovery is a story of continuous loops of chemical reprocessing. It looks a little complicated, but it's this process that makes the kraft pulping process viable.

2.5 SUMMARY

This lesson describes how pulp mills convert wood into pulp by chemical or mechanical processes or a combination of the two. Mechanical methods currently produce about 30% of the pulp produced in North America. They are the fastest-growing pulping technologies because of the economic advantages associated with high yields and because these processes are generally environmentally friendly.

Because mechanical pulps are relatively weak and relatively unstable (they yellow with exposure to light over time), pulp may also be produced using chemical processes. There are two basic chemical processes, the kraft process (alkaline) and the sulfite process (acidic). Over 90% of the chemical pulp in North America is produced using the kraft process because it produces a better-quality, stronger pulp and lends itself to chemical recovery to provide optimum economic and environmental benefits.

Chemically-produced pulps are low-yield (about half that of mechanically produced pulps) but have good strength properties and are very stable.

The properties and end uses for mechanical and chemical pulps are very different and are summarized in Table 1.

Table 1: *Properties and end uses for mechanical and chemical pulps*

Mechanical Pulp	***Chemical Pulp***	***Semi-chemical Pulp***
High yield (90+%)	Low yield (45%)	Intermediate yield (55%)
Short, impure fibers	Long, pure fibers	Intermediate properties
Weak	strong	
Unstable	stable	
Bulky, prints easily	Low bulk, harder to print	Intermediate properties
Processes		
Stone groundwood	Sulfite pulping	Neutral Sulfite
Refiner mechanical	Kraft pulping	
Uses		
Newsprint, phonebooks	Books, annual reports	Corrugating medium

2.6 REVIEW QUESTIONS

You will know that you have understood the materials in this module if you can easily answer the following simple questions:

1. What are the two basic process used to isolate fiber from wood?
2. What are the two kinds of mechanical pulping processes?
3. What are the two chemical pulping processes?
4. What are the advantages and disadvantages of mechanical pulping?
5. What are the advantages and disadvantages of chemical pulping?
6. What is Semi-chemical pulping and why is it used?
7. What are some typical products that use mechanically produced pulps?
8. What are some typical products that use chemically produced pulps?
9. What is the biggest use of Semi-chemical pulp?

Chapter 3

WOOD PREPARATION

OBJECTIVES

At the conclusion of this section, the student will be able to:

- ***Identify the major operations and equipment used in receiving and storing roundwood and chips.***
- ***Describe the chipping process.***
- ***Identify how chip screening improves pulp uniformity.***
- ***Explain how chips are stored and transferred.***
- ***Show why chip testing is necessary to ensure chip quality.***

3.1 KEY TERMS

Classifier - a piece of test equipment used to measure chip size distributions.

Chipper - machinery that turns logs into small chips of wood for pulping.

Cogeneration - the use of high-pressure steam in a turbogenerator to produce both electricity and lower-pressure steam.

Debarking Drum - machinery that removes the bark from logs.

Digester - a vessel that is used to pulp fibrous raw materials such as wood chips using chemicals.

Extractives - non-fibrous materials such as pitch, resins, and oils which make up 3% to 10% of wood.

Fines - small particles of wood that are too small to use in the pulping operation.

Hardwood - trees such as birch, maple, oak, eucalyptus, and gum which are characterized by broad leaves and short, slender fibers.

Knots - undercooked lumps of wood found in pulp after the pulping process, which tend to break up to form shives. Caused by oversized chips or especially dense wood.

Longwood - trees that are delivered to the pulp mill in lengths of 4 to 5 meters.

Mechanical pulp - pulp from a high-yield pulping process which depends primarily on mechanical action to separate fibers with minimal removal of lignin.

Oversize - chips which are too large to cook properly in the pulping operation.

Roundwood - sections of whole tree trunks which are harvested for lumber or pulpwood.

Residuals - leftover pieces of wood generated in the production of lumber and plywood. Residuals are converted into chips and sent to the pulp mill.

Shives - small splinters of undercooked wood remaining in the pulp after the pulping process. See knots.

Shortwood - wood which has been cut into shorter lengths of 1.5 to 2.5 meters before delivery to the pulp mill. Generally being phased out in most pulp mill operations in favor of chips, longwood, or tree-length logs.

Slasher - a type of saw that cuts longer logs into shortwood for older woodyards that cannot convey longwood.

Softwood - wood from coniferous trees characterized by needle-like leaves, such as pine, spruce, and hemlock. The fibers in softwoods are longer than hardwood fibers and make stronger paper.

Veneer - thin sheets of wood which are glued together to make plywood or used as a decorative outer surface for furniture and other wood products.

Wastewater - any excess of used water which is discharged from an industrial process or municipality.

Woodyard - the area of a pulp mill that receives chips and roundwood and processes the wood for use in the pulping process.

3.2 INTRODUCTION

The function of the woodyard and woodroom operations is to process raw material coming from the forest as logs into a form usable in the pulp mill, and to ensure that quality specifications are met. The first step in the process is to receive roundwood that will be stored, debarked, and chipped. Woodyards may also receive chips from satellite chip mills and wood residuals from lumber and plywood plants. Woodyard operations differ depending upon the geographic location and the type of pulp being produced.

3.3 RECEIVING AND STORING WOOD

Most modern woodyards receive long logs called longwood or tree-length roundwood. These logs are unloaded by large overhead cranes and sometimes by smaller mobile log-handling equipment. After unloading, the roundwood is stored until the mill is ready to convert it to chips. Older woodyards use shorter logs called shortwood. Since tree harvesting operations have shifted to more economical, long-length logs and even tree-length harvesting, shortwood systems have been almost completely eliminated.

Generally, roundwood is sorted as the log trucks are loaded in the woods. On arrival at the mill, the trucks are unloaded and the logs stored to keep wood species separated. A simple but very important sorting operation is separating the hardwood from the softwood. This is necessary because there are big differences between hardwood and softwood and the way each is pulped. Incoming wood may also be sorted by tree species, tree age and size, or type of pulp being produced. The climatic conditions around the mill can have a big influence in the woodyard. If there is a wet season when logging cannot be done or a dry season when fires occur, more wood must be stored to get through these times of short supply.

Roundwood storage can change many wood properties. Moisture loss is the first thing that occurs. Drying rates depend on the size of the logs and seasonal conditions. Small logs dry more quickly during hot summers. Dry wood breaks more easily when handled, and small chunks of wood can't be easily used. Stored wood is also more susceptible to insect damage. Wood resin and pitch content decrease during storage, and as with moisture, the rate of loss is fastest in small logs during the summer. Consequently, the pulp mill will recover fewer extractives that can be sold as turpentine and tall oil. Wood decay can occur if logs are stored for over a year, especially with hardwood.

Fungus growth can also be a problem for some pulp mills. Blue-stain fungi can discolor many types of softwood, reducing the brightness of mechanical pulps and increasing bleaching costs dramatically. Some mills use sprinkler systems to decrease the rate of growth of blue-stain fungi during warm, humid seasons, but any runoff has to be routed to a waste treatment system and reported in the mill's wastewater permit.

The conventional flow sequence for longwood involves transportation to the slasher deck, where long logs are reduced to manageable lengths with fixed circular saws mounted under a chain-fed log deck. More modern woodyards are built in a straight line so that 20-

to 30-foot long logs or tree-length stems 60 feet in length can be conveyed straight through with no turns from the log deck after storage, through the debarking drum, and into the chipper.

Very few mills rely on roundwood alone to supply the wood needed for the pulp mill. Most mills use from 25% to 100% purchased chips that are delivered from sawmills, plywood plants, satellite chip mills, and woods chippers. Trees have their greatest value if they are turned into lumber, but trees that are too small or defective are sent to the chip mill or pulp mill roundwood systems. Some logs are rejected at sawmills due to defects that can't be seen in the woods, or because the logs are too crooked, or broken in handling. These rejected logs are chipped at the sawmill or sent to the pulp mill for chipping.

In the process of lumber and veneer production, 50% to 70% of the incoming logs are converted into lumber and veneer. The residuals are made into chips that are transported to pulp mills. A few pulp mills use sawdust to make pulp, but most sawdust and bark is burned at the sawmill or plywood plant, sold to cogeneration power plants, or shipped to pulp mills for boiler fuel.

In summary, roundwood arrives by rail or truck. It is unloaded, sorted, and stored to minimize deterioration. Longwood may go to the slasher deck for cutting into prescribed lengths. From first-in/first-out storage, the logs are sent to the debarker and chipper.

3.4 DEBARKING

The next step in the operation is the removal of bark from the roundwood. If bark remains on the wood, pulping and bleaching are more expensive and the resulting pulp will have dirt specks. Bark from most pulpwood species, as shown in Figure 1, is dark in color, has high extractives content and also large quantities of sand and rock.

Figure 1: Bark

Several types of debarking methods and equipment are used in pulp woodrooms. These apply mechanical force to remove bark. The purpose of the debarker is to remove bark from roundwood with the minimum amount of log damage and loss of good wood.

Debarking drums (Figure 2) found in pulp mills and large chip mills range from 16 to 20 feet in diameter and 80 to 200 feet long. For optimum production, drums are filled about 50% full. They rotate from five to seven revolutions per minute. Logs are fed into one end and move down the drum to the discharge end. The force of logs hitting each other and the shell of the drum removes the bark. The speed of the logs moving through the drum is controlled by a movable gate at the discharge end and by the feed rate into the drum. As the bark is knocked off the logs, it falls through slots in the drum shell and onto a conveyor running the length of the drum.

Debarking waste is reduced in size by knife or hammer hogs. It can be pressed to remove water and burned in a waste fuel steam boiler. It can also be sold for mulch and other landscaping uses. Typically, a drum debarker removes about 90-95% of the bark from the logs. Bark removal depends on a number of factors, including wood species, season, and length of storage time. In winter, the dormant season for trees, bark is twice as hard to remove as in the spring and summer. When logs dry in storage, bark bonds more tightly to the wood. During the active growing season, some hardwood bark falls off logs in large pieces that require special handling to keep them from going with the logs to the chippers.

Figure 2: Drum debarker

Some reduction in winter bark adhesion can be accomplished by heating logs in hot water or steam. However, the rate of heat penetration in bark and wood is slow and often not enough to make debarking easy. If there is a source of low-pressure steam or hot water from the pulp mill, heating the logs before debarking can be cost-effective.

Ring debarkers are mechanical debarking machines found in sawmills. Sharp debarking tools mounted on a rotating ring are pressed against logs as they are fed one at a time through the debarker. The ring debarker is versatile, allowing feed and rotor speeds and pressures to be adjusted for a variety of sizes and diameters of logs.

Cutterhead debarkers (Figure 3) are another type of mechanical debarker. Instead of the ring of knives, cylindrical heads with teeth or spikes are pressed against the log. More wood is lost with cutterhead debarkers, but small logs can be debarked without too much breakage. Cutterheads are effective for frozen logs and species with high bark adhesion, but at the cost of high fiber loss.

All debarking systems cause some wood loss. The normal drum debarker wood loss is under 3%. When logs are held in a drum too long or mechanical debarking tool forces are too high, as much as 10-15% good wood loss can occur. Debarker operators must control debarking systems to maximize bark removal and minimize wood loss.

Any bark that is not removed in debarking is a contaminant. Bark has high pitch levels that form yellow to dark brown dirt spots in paper. Sand and dirt stick to bark in the woods and during storage at the pulp mill. These abrasive particles cause extra wear on equipment. More bleaching chemicals and pulp cleaning are needed to control bark-related dirt and to avoid abrasive wear.

In summary, bark is a contaminant that can increase the cost of making paper and negatively affect its appearance. The bark is removed from the logs using debarkers that break the bond between wood and bark. How well the debarker works depends on season, weather conditions, type of wood, and how much wood loss can be tolerated.

3.5 CHIPPING

Debarked logs must be reduced to small, uniform pieces for efficient use in the pulping process. Because wood varies in growth rate, grain direction, and strength properties, producing identical chips is not possible. Chip screening segregates oversize chips and fine particles so that the accepted chips are reasonably similar in size for pulping. Chips that are produced in the mill and those delivered from outside vendors are stored in outside chip piles so they can be conveyed to the digesters as the production rate demands. Chip quality is critical to digester operation and pulp quality, so chip uniformity is regularly monitored.

The most common chip size is twenty to twenty-five millimeters long by twenty to thirty millimeters wide and four to six millimeters thick, or roughly one inch square by less than ¼" thick. Chip thickness is the most critical dimension because it is the limiting factor in getting the cooking chemicals to penetrate the entire chip. Chips that meet these size

Figure 3: Cutterhead debarker

specifications are called "accepts". Chips that are too long and too thick are called "oversize chips". Chips that are the proper length but too thin or too narrow are called "pin chips". Fines are the smallest pieces of wood, including slivers, small chunks, and sawdust from wood products mills. Accepts, oversize, and pin chips are shown in Figure 4.

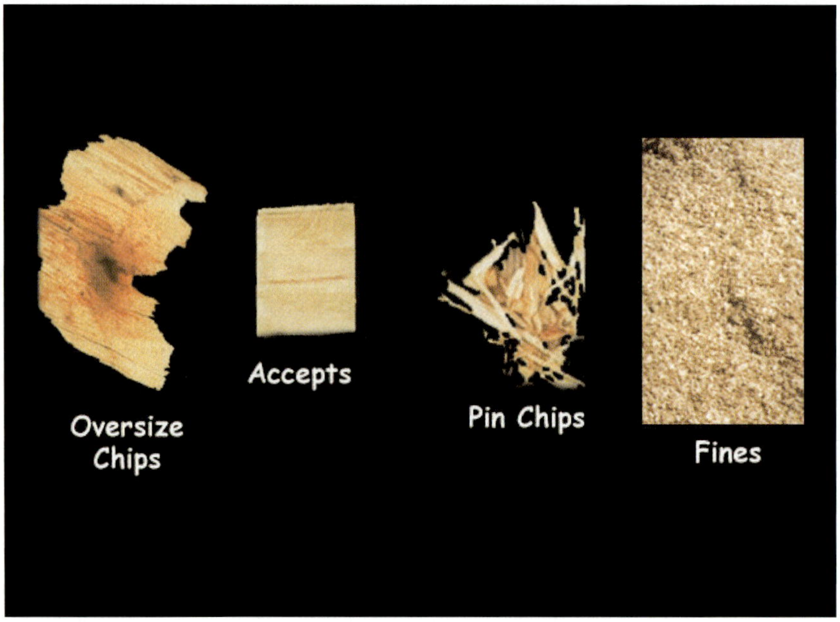

Figure 4: Accepts oversize and pin chips

To make uniform chips at high production rates requires chipping equipment with close tolerances between the knife and other chipper parts, constant and controlled rotation speed, and routine precision maintenance. If the equipment is not working properly, the chips will not be of good quality.

Logs are transformed by the chipper into smaller pieces of wood. There are a number of different types of wood chippers, but the disc chipper is most common. Disc chippers (Figure 5) have diameters ranging from 4 to 10 feet. Pulp mill woodyards use larger sizes for logs, and sawmills use smaller ones for slabs, trim, and reject lumber.

Disc chippers have four to twenty removable knives mounted radially (Figure 6). Located under each knife is a slot that runs the length of the knife. As the knives cut chips, the chips go through the slot. After passing through the disc, in one configuration, they are carried up by blades mounted on the edge of the disc and are blown out of the hood surrounding the disc. This is called an overhead discharge chipper. It the chips are allowed to fall

Figure 5: Disc chipper

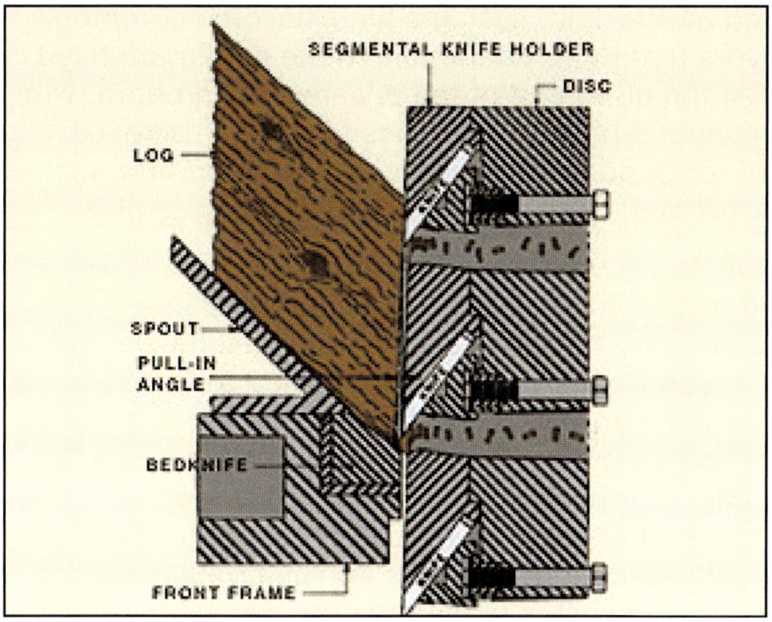

Figure 6: Disc chipper knife

onto a conveyor under the chipper, this is a bottom discharge chipper. The chips can also be transported away from the chipper by extending the hood so the chips can fall more gently onto the conveyor under the chipper. This prevents the chips from being damaged as they collide with the hood at high speed.

Another type of chipper, a drum chipper, is used for smaller wood pieces and residuals. A drum chipper has small knives mounted in the drum surface. Small logs and irregular chunks of wood such as log trim ends, broken log segments, and other clean, rock-free wood pieces, are dropped into a hopper above the drum. The knives cut out chip-like particles that are acceptable for most pulp mills. V-drum chippers have two cones attached at their apex. Four to ten knives are mounted, as in a disc chipper, from the apex to the base on each side of the V-drum, and logs are fed into the V. These are rugged chippers and have been successfully used for the large logs found in tropical forests and some areas of North America. While designs differ, all chippers work on the same basic principle. The wood is fed into the chipper infeed spout and drawn into rotating knives. The infeed spout angle, disc speed, knife to wood orientation, and method of discharge are some of the variables that determine the size and uniformity of chips. Logs are fed horizontally or by gravity into the chipper, and the disc can be at a right angle to the logs or tilted at about 60 degrees.

Not all pulp mills make their own chips. Chips can be purchased from satellite chip mills, field chippers with debarkers, or most commonly from sawmills and plywood plants. Chips may be transported to the pulp mill by railcars or barge, but trucks are the most common means of delivering chips. Moisture content of samples taken from each truckload is tested to estimate the amount of dry wood in the truck. Chip trucks are emptied in a quick and efficient way. They are turned on end! Rail cars can even be turned upside down.

3.6 CHIP STORAGE

Outside storage of chips has been practiced since the 1950s. Some mills with on-site chipping capacity can operate with less than one week of chip inventory. Most mills don't want to risk running out of chips, so they carry a higher inventory. Some older chip storage systems require several chip dozers to move chips from the delivery point out to the main pile and then back to the chip reclaim pit. However, this use of dozers breaks a lot of chips, and it's hard to rotate the inventory properly.

Since the 1980s, automated chip outstocking and reclaim systems have become common. Chips are put on the pile using belt conveyors that build up the pile to the maximum fixed height and keep moving in one direction. At the other end of the pile, a reclaim boom with blades scrapes chips off the pile onto a conveyor which usually goes to the chip screens. This design meets the primary requirement for good chip pile rotation. First-in / first-out operation ensures that chips are not left in storage so long that deterioration takes place.

It is also important not to allow the chip piles to become too compacted, which can happen if the piles become too high. Biological action results in an increase in temperature, much as a pile of grass or leaf mulch heats up in the center. However, chip pile heating should never be an issue if piles are limited in height and bulldozer usage is limited to the essential activity to outstock and reclaim. Chips are moved around the mill in several ways. Pneumatic conveying systems use blowers and pipelines to transport chips over long and short distances. Pneumatic systems have high operating costs due to their power consumption and can cause significant chip damage as chips are blown around turns in the pipeline. Belt conveyers are used for both short and long distances and can transport the chips without damage. While the initial equipment cost is higher, maintenance and power costs are lower. Chain and screw conveyors are used to move chips short distances such as out of reclaim pits, between hoppers and chip screens, and from under chippers. Bucket elevators are used to move chips vertically when the space between equipment is very limited.

3.7 CHIP SCREENING

Once chips are reclaimed from storage, they must be screened. Whole log chips are usually close to the target size distribution, but sawmill residuals have a wider size distribution due to chipping of irregular pieces of wood. Any chip handling will create some fine particles, depending on the intensity of the force on the chips during conveying and dozing. Oversized chips called oversize and small pieces called fines need to be removed to allow for uniform pulping of the remaining chips. The oversize are reprocessed, but the fines go to wood- and bark-fired boilers, sawdust digesters, or particleboard mills.

Before we can discuss chip screening in detail, we need to understand the size classes in a chip distribution. As chip screens have improved, so have the classifiers used to measure chip size distributions. The Williams Classifier, which served the industry for decades, effectively measures chip length and width but not thickness. Pulping researchers then found that chip thickness is the key chip dimension for kraft pulping. A new chip classification system was then developed to measure the chip fractions that really matter in digester operation. This classifier design is used as a standard in some countries and is widely used in North America.

As the photo shows (Figure 7), this classifier has four trays and a pan at the bottom. A sample of chips placed in the classifier will be segregated into oversize on the 45 mm round hole, overthick on the 8 or 10 mm equally spaced parallel bars, acceptable chips on the 7 mm round hole screen, pin chips on the 3 mm round hole screen, and fines on the bottom

Figure 7: Chip classifier

tray. All new chip classifiers separate out overthick chips as well as the standard classifications that have been used for decades.

Long and wide oversize chips can cause handling problems in conveying and feeding. Oversize chips are almost always overthick, too. Pulping chemicals cannot completely penetrate overthick chips during the pulping process. After the chips have been cooked, the hard, uncooked centers of overthick chips, called pulp knots, remain in the good pulp. They are easy to screen out of the pulp, and most knots are repulped, but this results in lower digester capacity and yield. Some undercooked fiber layers of the knots break off as the pulp is pumped from the digester. These fiber bundles are called shives. Pulp mills must screen them out of the pulp before it goes to the bleach plant or the paper machine. For these reasons, it is far better to screen chips before cooking than try to handle the resulting problems after pulping is complete.

Chip screening has made significant advances in the last fifty years. The first screens installed in the 1960s had a gyrating action that fed chips across wire mesh or punched plates. The idea came from the material handling industry that screens rock and coal particles. The top deck has round or square openings from 1-1/2 to 3 inches wide, and the bottom deck has 1/8 to 3/8 inch wide wire mesh or punched plates. This type of screen is still used in wood products mills and in pulp mills, sometimes in combination with other screen types specifically designed for overthick chips.

The disc screen was developed to remove overthick chips more effectively than a gyrating screen. It consists of parallel discs precisely mounted on a series of shafts that rotate in the same direction. The disc spacing is usually between 7 and 10 mm. Overthick chips cannot pass between the discs and are carried off the end of the screen. Once overthick chips are screened out, they are either sliced or crushed to make them thinner or open up the wood structure so that pulping chemicals can penetrate more uniformly into the center. They can then be put into the digester with the rest of the chips.

Chips can become contaminated with materials that harm pulping equipment and reduce pulp quality. Ferrous metal contaminants are typically parts that fall or break off equipment during chip making or transporting. Maintenance materials such as nails, bolts, nuts, fasteners, and welding rods are also commonly found. Most of this iron material can be removed by one or more strong electromagnets that are mounted over chip conveyors and pull metal out of the chips flowing under the magnet.

Small quantities of aluminum cans and log tags are not a problem in kraft mills, since they dissolve during pulping. Plastic is the worst contaminant in chips. It gets into chips mainly by careless disposal of packaging materials and food wrappers. The best plastic contamination reduction program is continuous and effective education of everyone about the seriousness of plastic contamination. Plastic contamination is very difficult and expensive to remove and can produce defects in the final paper. Rocks and sand are very damaging to mechanical pulping systems because these dense materials can wear and break refining equipment. Chip washers are used in some mills to remove hard abrasive materials.

Chip quality control testing is important to determine whether the chips meet the needs of the pulp mill. The basic tests measure moisture content, size distribution, and level of contaminants. Moisture content is needed to convert green weight of chips to dry weight and to calculate pulping chemical requirements in digesters. Bark content is the primary contaminant test. Periodically, a mill will run wood specific gravity, bulk density, and species analysis to determine whether wood properties are changing over time.

This concludes the lesson on chip production and handling. This lesson identified the equipment and processes involved in chipping wood for pulp mills. Wood products mills and satellite chip mills are supplying an increasing proportion of the chips used in pulp mills. The interactions between chipping, chip screening, contaminant removal, and chip storage were discussed. Chip quality control programs measure the efficiency of chip production systems. Uniform, high-quality chips free of contaminants are the first step in producing high-quality pulp for papermaking.

3.8 REVIEW QUESTIONS

You will know that you have understood the materials in this chapter if you can easily answer the following questions:

1. Explain why the woodyard separates hardwood and softwood chips.
2. Why are mills using more purchased chips?
3. Explain how a debarking drum works.
4. List the impacts that longer storage time has on wood.
5. What happens to the bark that is removed?

Chapter 4

KRAFT PULPING

OBJECTIVES

After completing this section, the student will be able to identify and explain:

- ***How wood chips are converted to kraft pulp.***
- ***Equipment used in kraft pulp making.***
- ***Some of the variables which control the delignification process.***

4.1 KEY TERMS

Black Liquor - The spent cooking liquid from the kraft pulping process.

Brownstock - The pulp slurry after washing but before any bleaching process. It is brown because of the remaining lignin in the fibers.

Cellulose - The main structural constituent of woody plants. Wood cellulose is the material remaining after much of the lignin and certain carbohydrates have been removed from wood cell walls by pulping and bleaching.

Chemical Pulping - A process to dissolve lignin by the use of chemicals in order to liberate the cellulose fibers.

Chlorine - Cl_2. A bleaching chemical that removes the colored lignin in brownstock.

Clarifier - A type of tank used in chemical recovery to clarify green or white liquor by allowing entrained particles to settle out of the main upward flow and be removed as slurry from the bottom of the tank.

Concentrator - An evaporator supplied with fresh steam, used in the final stage of concentration of an aqueous solution.

Consistency - The percentage by weight of ovendry fibrous material in a stock or stock suspension. It is sometimes called density, solids, or concentration.

Cooking Liquor - The solution of chemicals used in pulping to remove the lignin. Depending upon the pulping stage, the liquor may be white (fresh kraft cooking liquor), black (after pulping), or green (after chemical recovery but before causticizing).

Digester - A vessel that is used to pulp fibrous raw materials such as wood chips by chemical means.

Dioxin - The common term for a family of chlorinated organic compounds that are mildly to highly toxic. Can be produced in the bleaching process if chlorine is present.

Disc Refiners - A piece of rotating equipment with two or more discs which have bars on their surfaces. The discs are set close together so that wood chips that pass between the two surfaces are separated into fibers.

Evaporator - Equipment to remove water from an aqueous solution (e.g. black liquor) to produce a solution of higher concentration.

Groundwood - a high-yield mechanical pulp produced by pressing logs sideways against a rotating pulp stone.

Hemicellulose - a broad class of polysaccharides which are associated with cellulose in plant cell walls. Hemicelluloses have a lower molecular weight than cellulose.

Kraft Pulping - an alkaline pulping process that uses sodium hydroxide and sodium sulfide to dissolve lignin and release cellulose fibers.

Lignin - a substance that acts as a glue between cellulose fibers to form the woody structure of plants.

Lime Mud - Slurry or cake of fine particles of calcium carbonate derived from the reaction of reburned or fresh lime with green liquor.

Mechanical Pulping - A high-yield pulping process which depends primarily on mechanical action to separate fibers with minimal removal of lignin.

Pulping -The process of separating a cellulose-containing raw material, such as wood or straw, into individual fibers which can be reformed to make paper.

Recausticizing - A part of the kraft recovery process in which lime is added to green liquor to form white liquor.

Slaker - An agitated tank in which lime is reacted with green liquor or water to form calcium hydroxide.

4.2 INTRODUCTION

Kraft pulping is the world's number one process for making pulp from wood. Kraft pulping was discovered during a search for a makeup chemical to provide sodium for soda pulping. In 1879, Dahl discovered that the use of sodium sulfate as a makeup chemical produced a cooking liquor that pulped the wood faster and gave a stronger pulp. The strength of the pulp led to the name "kraft", since "kraft" means "strength" in German. Kraft pulping is a fully chemical process. The chemicals can be continuously recovered and reused, all the while producing steam and electricity to run the mill operations. The kraft process can be used successfully on virtually any lignocellulosic material, including almost any species of wood, as well as non-wood materials such as bamboo and sagebrush. All of these factors have contributed to the rise of kraft pulping, which now produces 70-80% of the pulp used in North America.

4.3 PULPING CHEMISTRY

Wood consists mainly of cellulose, hemicelluloses, and lignin, all intermixed in the form of parallel, hollow tubes we call fibers. Lignin content varies by wood species, with more lignin in softwoods than in hardwoods. The lignin in the wood holds the fiber structure together and has color. In order to separate the fibers and produce a white pulp, the lignin must be removed. The chemical removal of lignin is called delignification. Some of the lignin is in the middle lamella region between fibers, where it glues them together (Figure 1). Most of the lignin, however, is in the fiber walls. This lignin is quite difficult to remove.

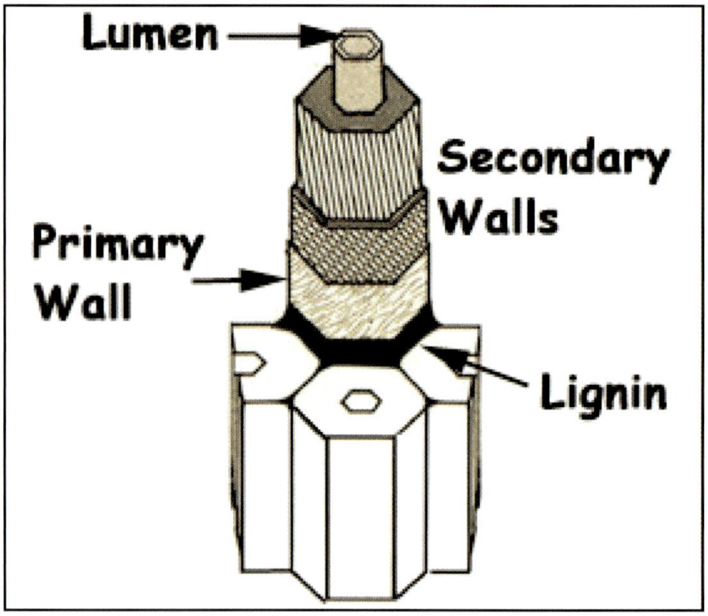

Figure 1: Cell wall

Kraft pulping is done with a mixture of sodium hydroxide and sodium sulfide in water. This solution is called white liquor. The hydroxide and sulfide ions are the active chemical species, which attack the lignin and gradually break it down into soluble fragments. Unintentionally, some degradation of cellulose and hemicelluloses also occurs. When pulping is complete, most of the sodium hydroxide has been consumed, along with about half of the sodium sulfide. The spent liquor has a very dark color due to its dissolved lignin content and is called black liquor.

The kappa number is the industry's standard measure of how much lignin remains in the unbleached pulp. The higher the kappa number, the higher the residual lignin content and the darker the pulp. Different grades of pulp are made to different target kappa numbers. For example, bleachable softwood pulp is typically in the kappa range from 25 to 35.

Bleachable hardwood pulp is between 12 and 18. Linerboard kappa numbers are much higher, usually about 90. The kappa number may be used to determine the percentage of lignin remaining in the pulp. Simply multiply the kappa number by 0.147 to obtain the percentage of lignin. For example, a bleachable softwood pulp with a kappa number of 30 would contain 30 x 0.147 = 4.41% lignin by weight. Considering that the softwood started out containing about 30% lignin by weight, we can see that the pulping process removes a large fraction of the lignin from the wood.

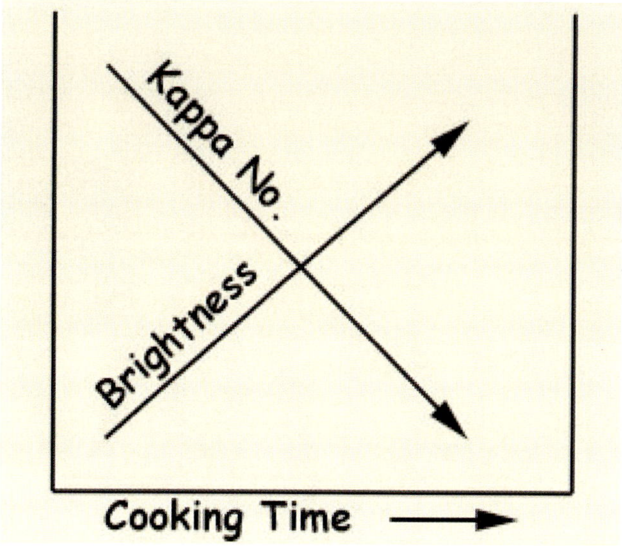

Figure 2: Relationship between kappa number and brightness

Why do mills cook to different kappa numbers? Primarily because pulp with a low kappa number (and therefore low lignin content) is brighter and easier to bleach (Figure 2). The pulp requires lower bleaching chemical application rates in the bleach plant, and so is less expensive to bring to high brightness. The downside of a low kappa number pulp is that the fibers may be weaker and the pulp yield is lower. When producing a grade of paper where high strength is important and a brown color is acceptable (such as linerboard), a higher kappa number pulp is usually better. When the whiteness of the pulp is important, a low kappa number will minimize the costs in the bleach plant.

4.4 PROCESS AND EQUIPMENT

A typical pulp mill is a very large facility, containing a collection of tanks, pumps, pipes, and specialized equipment designed to produce a high quality product. This section will focus on some of the specific equipment that is required to produce kraft pulp.

Originally, kraft pulping was done in batches in large pressure vessels called batch digesters. A batch digester cooks one batch of wood chips at a time. It is simply a big pressure cooker, using a set of ingredients (wood chips and cooking liquor) and a recipe (including time and temperature) to make a predictable product - kraft pulp.

The wood chips are fed into the digester from the top (Figure 3). Simultaneously, white liquor is added; the amount is based on the requirements for the grade of pulp being made and the species of wood used in the process. Some black liquor is also added to ensure that there is enough liquid to cover the chips. The amount of black liquor added to the digester is determined by the desired liquor-to-wood ratio (L/W). Typical batch operations have L/W of about 3.5-4.0 to provide enough liquor for circulation.

During impregnation and cooking, the liquor is usually circulated through the digester by extracting it through a screen, heating it, and pumping it back into the top and bottom of the digester. This helps to even out the chemical and heat gradients which inevitably occur inside the column of cooking chips. Generally, the entire liquid volume is cycled through

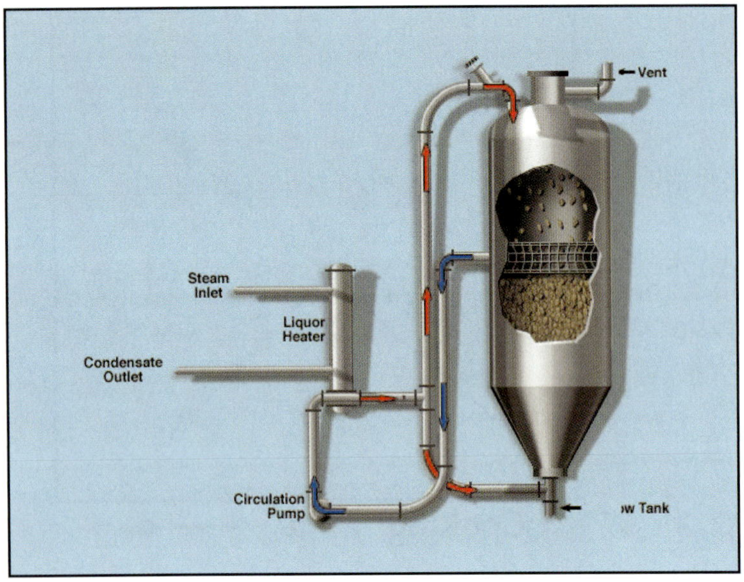

Figure 3: Feeding chips into a batch digester

about every 10-15 minutes. Older batch digesters are often heated by direct injection of steam instead, without any forced liquor circulation.

The liquor is circulated and heated until the desired operating temperature is reached, typically 160-170ºC. The cooking temperature is then maintained as constant as possible for a period of about one to two hours, depending on the type of pulp being produced and its target kappa number. Early in the cooking cycle, air and other non-condensable gases are removed through a special vent. At the end of cooking, the blow valve at the bottom of the digester is opened and the pressure release forces the cooked chips and black liquor out through the blow line and into the blow tank. A batch kraft mill has many digesters - from four to as many as fourteen - operating on a fixed production schedule.

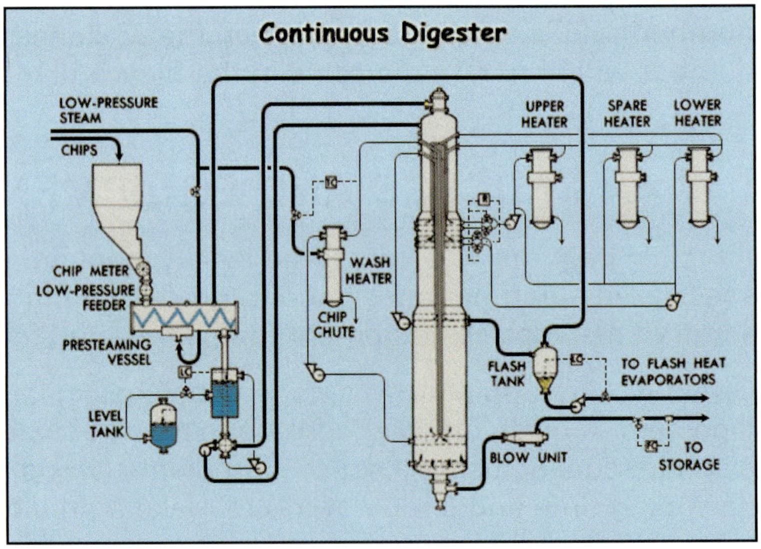

Figure 4: Continuous digester system for kraft cooking

Since the 1960s, continuous digesters have become increasingly popular, in some cases replacing batch digesters. Continuous digesters do just what the name implies - they make pulp continuously. Refer to Figure 4 for the description that follows. The wood chips are conveyed to a chip bin, from which a rotary feeder - the chip meter - transfers them at a known rate into a low-pressure feeder and then into a steaming vessel where air is driven out of the chips and they are heated. It is now common practice to apply steam in the chip bin as well. The steamed chips fall from the steaming vessel into the chip chute, where they first meet the white liquor. A high-pressure feeder then directs the chips and liquor to the top of the digester system.

The slurry of chips in liquor enters a cylindrical top separator. A rotating helical screw controls the rate at which the chips are delivered into the digester and allows some of the liquor to drain from the chips and return to the high-pressure feeder. A column of chips and liquor is created inside the digester and moves down slowly through the digester tube. Chips and white liquor are continuously added to the top, black liquor is removed near the midpoint, and pulp is removed from the bottom of the digester.

In order to produce a uniform pulp, the cooking liquor must saturate the chips completely, all the way to the chip centers. This saturation, often called impregnation, of white liquor into the chips is done under pressure at about 120ºC. No delignification happens at this point; the temperature isn't high enough. The location of the impregnation zone depends on the type of digester. In a single-vessel system, it is in the uppermost section of the digester. In a two-vessel system, liquor penetration takes place in a separate impregnation vessel.

Once impregnation is complete, the chips and liquor are heated to pulping temperature, usually 160-170ºC, and delignification begins. In a single-vessel digester, two consecutive liquor circulation loops are used; each extracts liquor out of the digester through screens, heats it in a heat exchanger, and pumps it back into the middle of the chip column via one of the concentric "downcomer" pipes. These pipes hang down in the center of the digester vessel. In a two-vessel system, all the heating is done in the bottom circulation loop, which transports impregnated chips and liquor from the bottom of the impregnation vessel to the top of the digester vessel and returns excess liquor. In both types of digesters, what follows is cooking time in one or more zones until there has been enough delignification to reach the target kappa number for the grade of pulp being made.

When pulping is complete, the hot spent liquor is removed through extraction screens, located near the digester midpoint, and sent to a pair of flash tanks in series. In these flash tanks, the pressure drop causes steam to flash from the hot liquor. The low-pressure steam from the first flash tank is used to pre-steam the chips in the steaming vessel mentioned earlier. The atmospheric steam from the second flash tank is used in the chip bin.

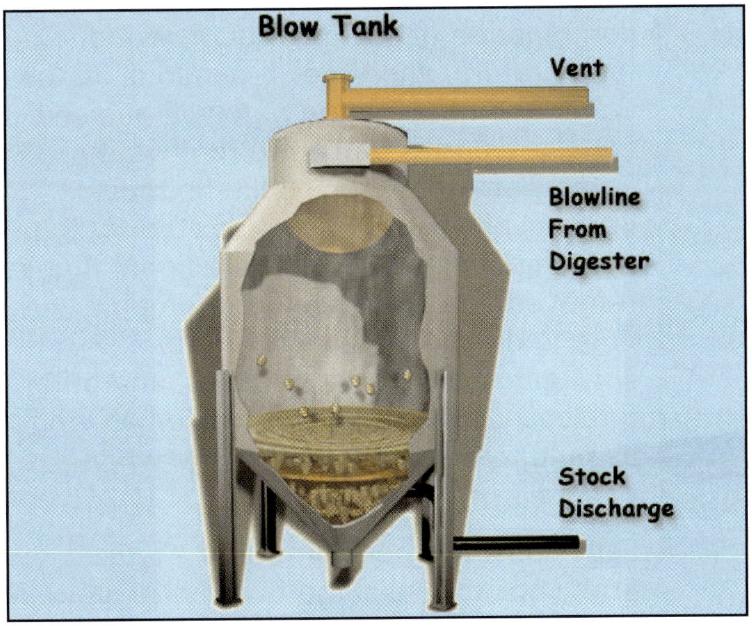

Figure 5: Blow tank

Near the bottom of the digester, the slowly rotating paddle arms of the outlet device scrape the cooked material off the bottom of the chip column. Washer filtrate, now called cold blow flow, is pumped in and dilutes the cooked material to a consistency that will allow the pulp to flow out of the digester. Enough excess filtrate is pumped into the bottom of the digester that some of the liquid moves upward through the chip column, toward the midpoint of the digester. The upward flow of this filtrate provides the first washing treatment for the pulp. The diluted pulp suspension is directed down into the outlet device and along the blowline to the blow tank or diffusion washer.

A blow tank is a large, unpressurized cylindrical vessel which receives the cooked material and spent liquor from a digester (Figure 5). The turbulence of the discharge flow transforms the cooked chips into pulp. A blow tank has an agitator that continuously mixes the pulp suspension. In a batch pulping process, some of the heat from the liquor is recovered in a blow heat accumulator and is generally used to heat process water needed elsewhere in the mill. Ideally, the pulp suspension in the blow tank consists of individual pulp fibers. In reality, some poorly cooked fiber bundles and larger "knots" are also present. The spent liquor is a solution of dissolved inorganics (the "spent" cooking chemicals) and dissolved organics, primarily from lignin.

Regardless of which type of digester is used, every pulping process requires an appropriate amount of liquor, wood, temperature, and time. These four factors are interdependent. The right amount of liquor, at the right strength, is necessary; it is usually referred to as either active or effective alkali, expressed as a percent applied on wood. Once in contact with the chips, the white liquor must thoroughly penetrate into them before delignification can take place. How fast the penetration occurs is controlled by chip thickness, alkali concentration, and temperature. Liquor temperature affects the rates of the delignification reactions that we want, as well as the degradation of hemicelluloses and cellulose, which we would like to minimize. Too much liquor, heat, or time will result in inferior pulp strength and low yield. Yield is the amount of acceptable pulp obtained as a percentage of the wood put into the digester.

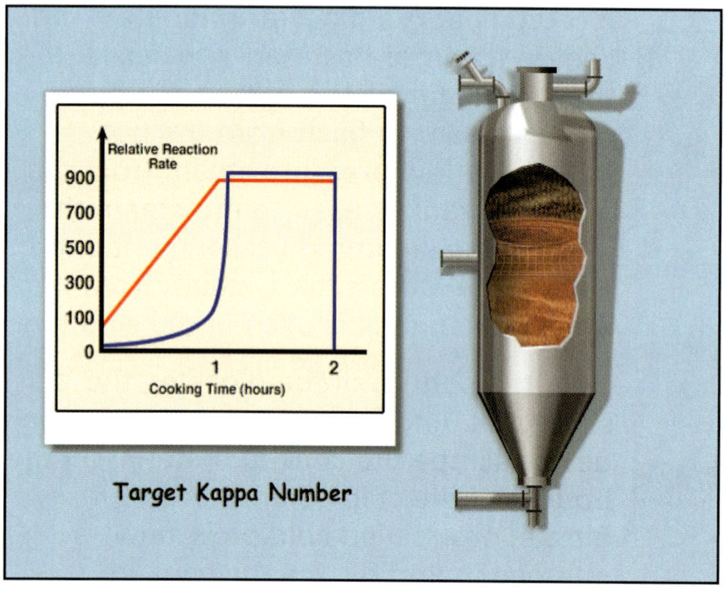

Figure 6: H-factor

What determines the appropriate combination of cooking liquor, wood, temperature, and time? It depends on the wood and on what the pulp will be used for. The wood is pulped to remove the lignin. The longer the cook, the more lignin is dissolved. Since the color of the pulp depends on it, the more lignin that's removed, the lighter-colored the pulp. Pulp to be used in linerboard for corrugated boxes, brown bags, and other unbleached papers is not cooked as long as pulp to be bleached to make white papers.

Let's consider the example of a typical kraft cook to make linerboard basestock in a batch digester. The target kappa number is 90. After filling the digester with the appropriate amounts of chips and liquor and sealing it, we raise the temperature to 170ºC by direct steaming for one hour. Then we cook the chips for one hour at maximum temperature. This is a relatively short kraft cook, but it represents the pulp grade requiring the least delignification.

Figure 6 is a chart showing both the temperature profile for this cook and the corresponding rate of delignification. Notice that as the temperature rises, the rate of delignification increases as well. When the temperature levels out, so does the delignification rate. The area under the curve of relative delignification rate versus cooking time is known as the H-factor. The H-factor combines the temperature and time of a kraft cook into a single variable.

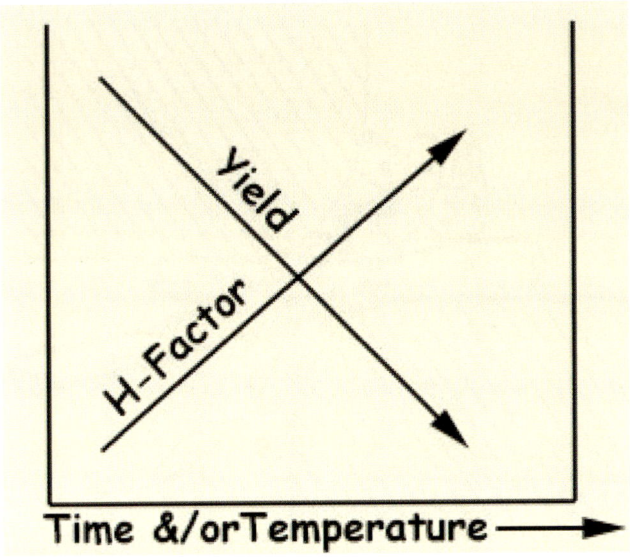

Figure 7: Relationship between H-factor and yield

Cooking to a constant H-factor should result in the same kappa number in the resulting pulp. H-factor is a parameter used in every kraft mill to correct for irregularities in temperature control. For example, if you are short of steam and the temperature is slow to increase, the H-factor will allow you to predict the extra time required to reach the target kappa number.

There are many variables involved in Kraft pulping. In general, the higher the H-factor, the lower the pulp yield (Figure 7). A longer cooking time, a higher cooking temperature, or a combination of both can lead to an increase in the H-factor. There are two reasons that higher H-factors result in a lower pulp yield. First, there is less lignin left in the pulp to contribute to total yield. Second, more of the cellulose and hemicellulose content has been degraded. Figure 8 shows how the yield decreases as lignin is removed during kraft pulping, and illustrates the main components of yield.

Here's an example calculation. We start with 100 kg of wood, composed of 45 kg of cellulose, 25 kg of lignin, 25 kg of hemicellulose, and 5 kg of extractives. A pulp yield of 50% means that after cooking, half of the wood remains as unbleached pulp. Its composition is 40 kg of cellulose, 5 kg of lignin, 5 kg of hemicelluloses, and no extractives. Thus, we have removed 80% of the original lignin. At the same time, all of the extractives have been dissolved, along with 80% of the hemicelluloses, but only a small amount of the cellulose. That's good news, since pulp and paper is a cellulose-based industry. Obviously, the chemical composition of the wood itself is an important variable in kraft pulping, and the higher its lignin content, the slower the pulping and the lower the pulp yield.

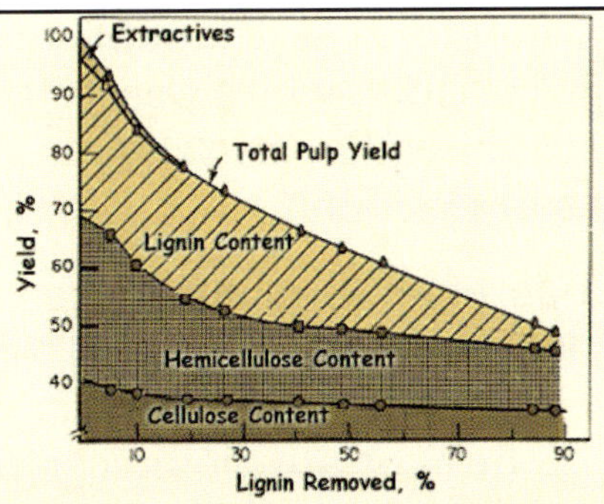

	ORIGINAL WOOD FIBER	PULP	% REMOVED
CELLULOSE	45 UNITS	40 UNITS	11
LIGNIN	25	5	80
HEMICELLULOSE	25	5	80
EXTRACTIVES	5	0	100
TOTAL	100	50	50

Aurell and Hartler

Figure 8: Relationship between yield and lignin removal

4.5 PULP DEFIBERIZING AND DEKNOTTING

Once pulp is in the blow tank, one or more operations may occur before it is washed. If the

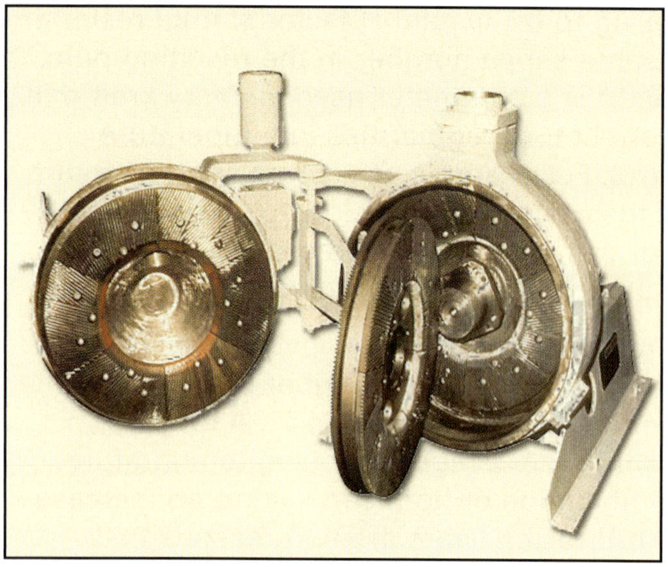

Figure 9: Disc refiner

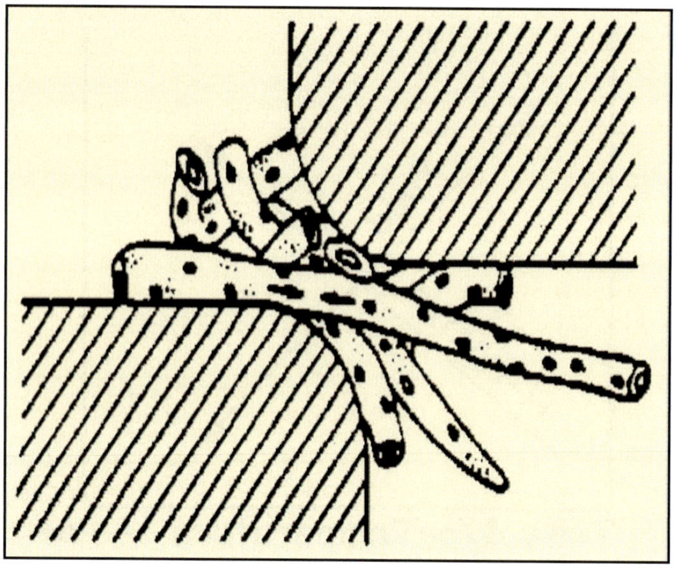

Figure 10: Pulp fibers passing between refiner plate bars

pulp is a high kappa grade, such as linerboard basestock, some mechanical action is necessary to separate all of the FIBERS. This "defiberizing" process is usually done in a special disc refiner such as the one shown in Figure 9. It has a central rotor with a set of refining plates on each side, and two sets of opposing plates attached to the stator. Such a machine is referred to as a "double disc" refiner, because of the two separate pockets for defiberizing. The pulp, which passes between the bars of the opposing refiner plates (Figure 10), is still suspended in black liquor, which makes defiberizing easier. The operation is termed hot stock refining.

A different operation is usually necessary with lower yield kraft pulps destined for bleaching. Wood chips that are too thick (that is, greater than about 10 mm) or are biological knots (where branches were once attached to the trunks of trees) won't cook completely, resulting in "knots" which must be removed from the pulp. The unwashed pulp is passed through a pressurized "knotter" (Figure 11) to separate this material. If the amount of rejected material is high, a secondary knotter may be needed as well. Typically, the knots are sent back to the digester.

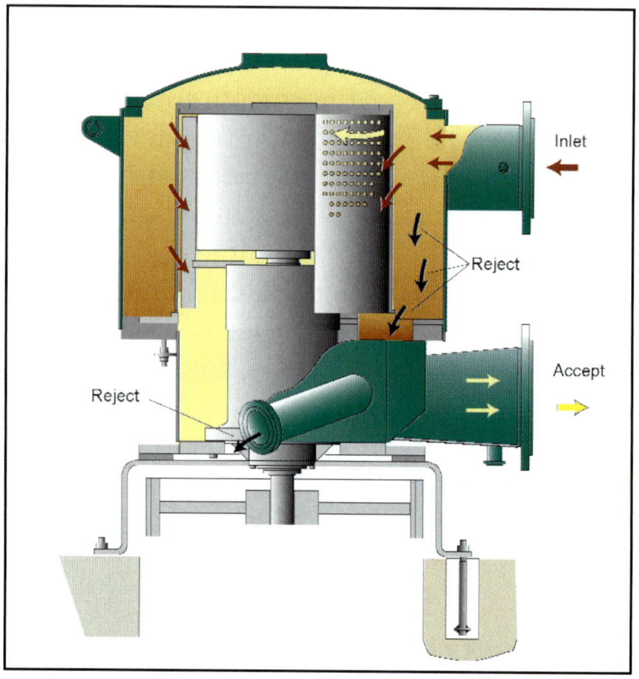

Figure 11: Pressure knotter

4.6 BROWNSTOCK WASHING

One of the key aspects of the kraft process is that the spent cooking chemicals can be continuously recovered, reactivated, and reused. Before this can happen, though, the dissolved organic and inorganic chemicals in the black liquor need to be separated from the pulp. This is done in brownstock washing. Black liquor is removed as wash filtrate and sent to the chemical recovery side of the mill to be concentrated and burned as a fuel; at the same time, its cooking chemicals are recovered.

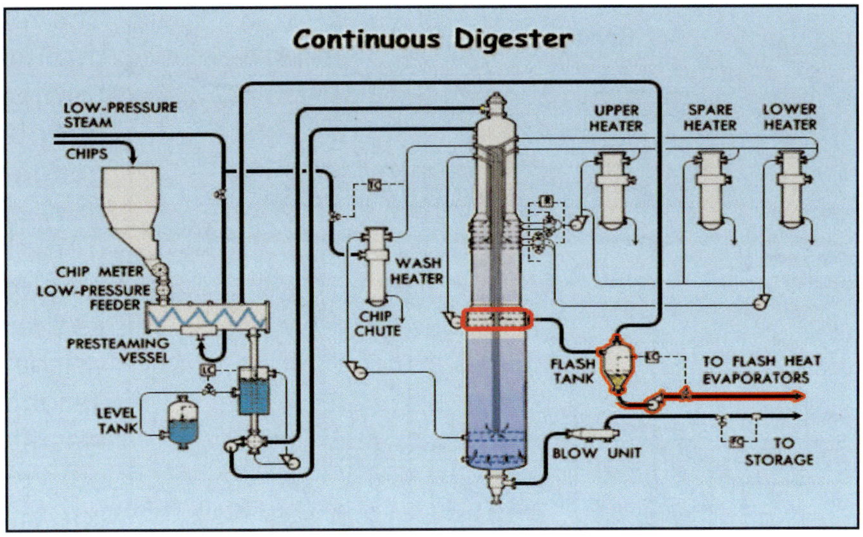

Figure 12: Washing inside the continuous digester

As shown in Figure 12, washing begins inside the continuous digester. As the column of cooked chips nears the bottom of the digester, black liquor is displaced upward by cooler brownstock washer filtrate injected from below. The displaced black liquor carries residual alkali upward. This liquor exits through extraction screens and the flash tanks and goes to the liquor recovery system. The pulp leaving the bottom of the digester proceeds to the next step in the process: brownstock washing.

To remove as much black liquor as possible, pulp from both continuous and batch digesters is washed on brownstock washers, often using a series of two to four rotary vacuum drum washers (Figures 13 and 14).

Figure 13: Photograph of drum washers

Washers use either vacuum or pressure to assist in removing liquor from the pulp. Each individual washer removes only a portion of the black liquor; therefore, the pulp passes over a series of such washers. Over the entire washer system, the removal of the dissolved chemicals in the black liquor will be very efficient.

A rotary vacuum washer has a wire-mesh-covered horizontal cylinder which rotates slowly in a vat containing the pulp slurry. A cross-section of a typical cylinder is shown in Figure 15. A vacuum is applied on the inside of the cylinder. As the cylinder rotates through the pulp slurry in the vat, the vacuum draws pulp onto the face wire, causing a pulp mat to form. As the cylinder rotates out of the vat, wash water is applied by showers, displacing the residual black liquor. Further rotation brings the cylinder to a point where the vacuum is cut off and the pulp mat is removed on a "doctor board". It leaves the washer via a repulper screw, is diluted to low consistency, and is pumped to the vat of the next washer.

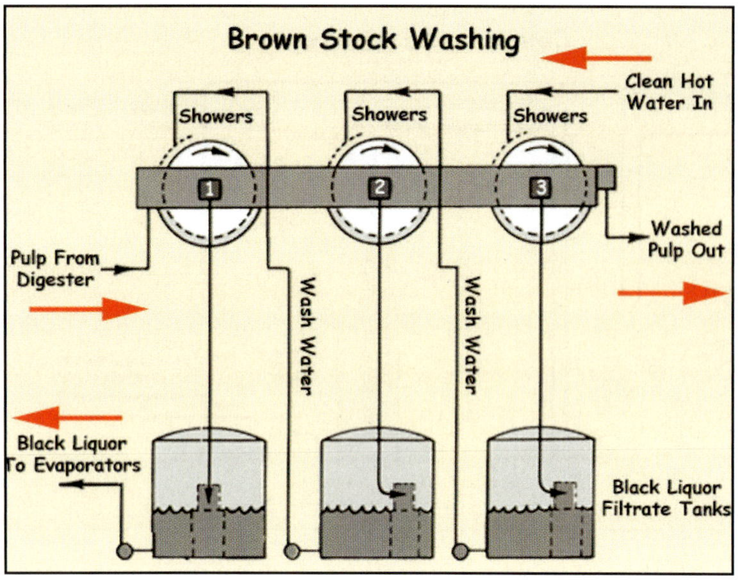

Figure 14: Drum washer diagram

The filtrate is removed from the cylinder through a piping arrangement called a "dropleg", going into a seal tank on the floor below. The vacuum in the drum is generated by the siphoning action of the flow of the filtrate in this dropleg. Filtrate from the third washer is used as the dilution water ahead of that washer and as the shower water for the second washer; the same is true between the second and first washers. So brownstock washing is a counter-current operation - pulp goes forward along the production line and wash water goes the other way (Figure 14). This arrangement maximizes the efficiency of pulp washing, because the cleanest pulp is washed with the cleanest water.

Pressure washers are organized in the same way; the only difference is that instead of a vacuum being pulled inside each cylinder, air pressure is applied inside the sealed hood surrounding the cylinder, eliminating the need for a long dropleg. Rotary pressure washers can have two or three displacement stages in one machine, allowing a single pressure washer to function as efficiently as a series of vacuum washers.

After continuous digesters, it is common to find diffusion washers. These devices are enclosed machines having concentric filtrate extraction screens between which medium-consistency pulp travels upward and wash water is injected into the pulp by slowly rotating nozzle arms (Figure 16). This is a very efficient form of brownstock washing; it avoids the repeated dilution of the pulp suspension to 1-2% consistency that happens in a series of vacuum drum washers. Atmospheric diffusers usually have two stages, piped for counter-current filtrate flow; some older single-stage diffusers can also be found. It is also possible to install a pressure diffuser between a digester and an atmospheric diffuser, increasing the overall efficiency of washing. Alternatively, atmospheric diffusion washers may be followed with two or three stages of washing on rotary vacuum washers or pressure washers.

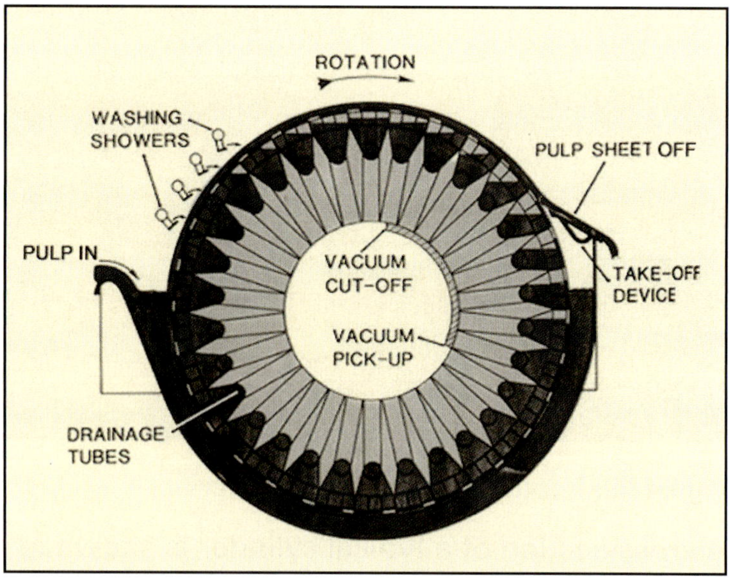

Figure 15: Cross-section of washer cylinder

4.7 PULP SCREENING

The pulp coming from brownstock washing is at medium consistency. The next step in the process is brownstock screening, at low consistency, which removes any oversize, troublesome, and unwanted material from good papermaking fibers. Screening equipment uses a perforated barrier which allows acceptable FIBERS to pass but rejects bigger particles. All screens use some sort of mechanism to clean the openings in the perforated barrier, preventing them from clogging.

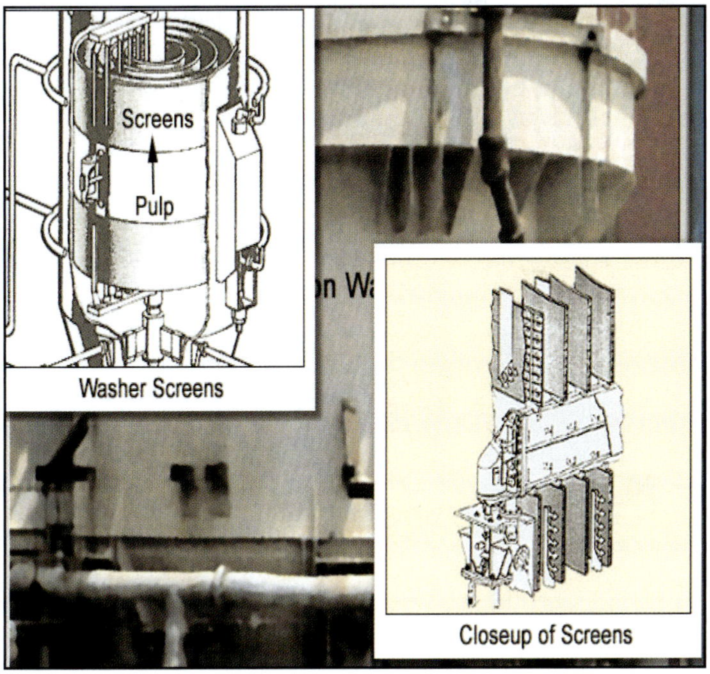

Figure 16: Diffusion washing

A pressure screen has a cylindrical body enclosing an inner perforated cylinder or screen basket Figure 17). Pressure forces the pulp slurry through the holes or slots in the screen basket as accepts; unwanted particles are removed as rejects. The most common element for cleaning such a screen is a rotating hydrofoil (Figure 18). As it passes by the openings in the screen, it produces pressure pulses which clear the openings of trapped debris. The trailing blade of the foil generates a slight vacuum, pulling fiber through the holes or slots in the screen basket. Pressure screens were originally used in paper mills, but their efficiency and versatility made them an essential part of pulp mills, too.

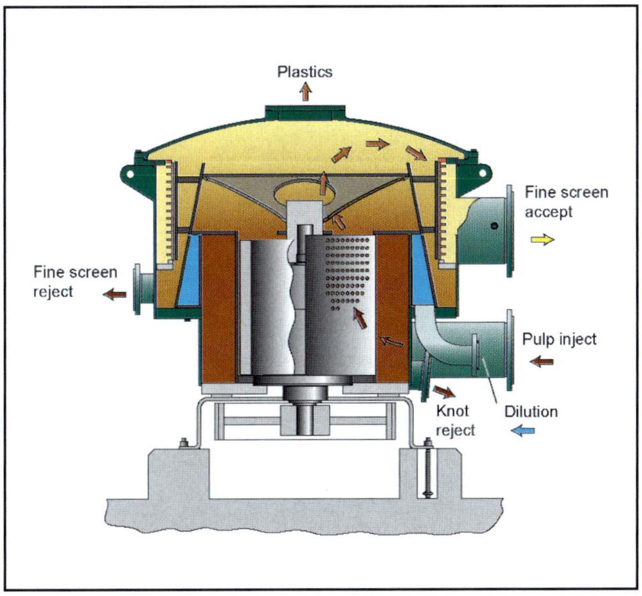

Figure 17: Pressure screen

Figure 18: Rotating hydrofoil

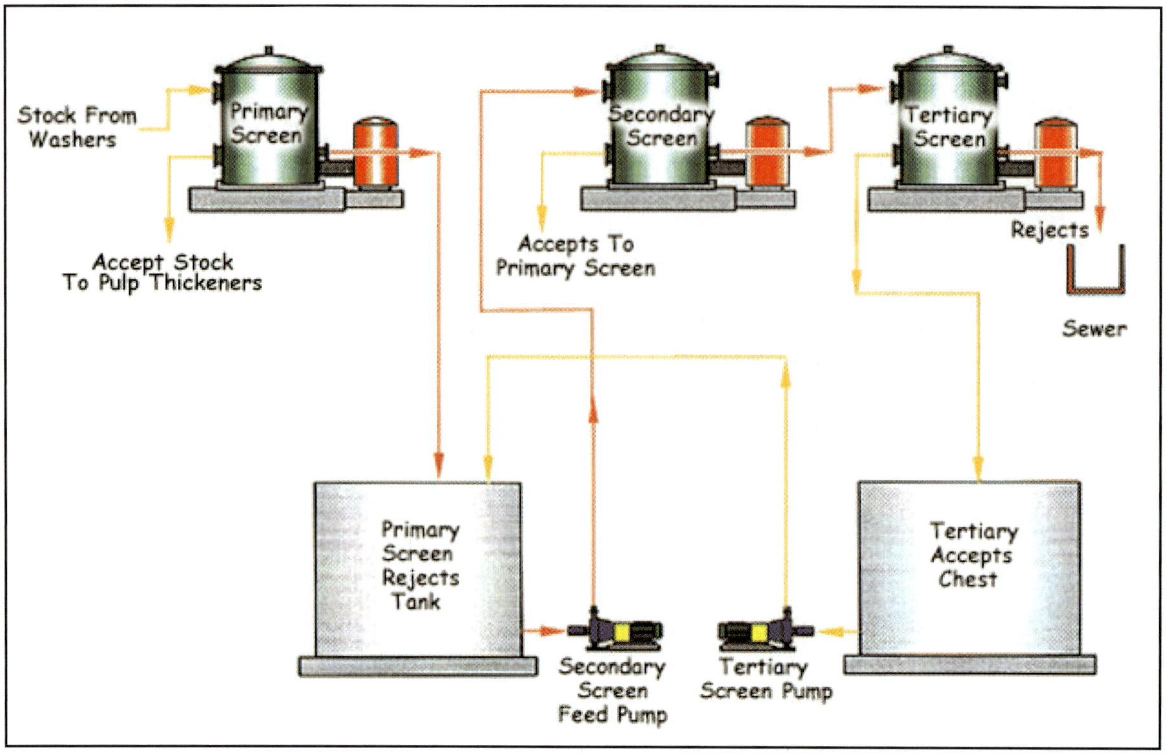

Figure 19: Pulp screening systems

Figure 20: Brownstock decker

Figure 21: High-density storage tank

As in most separation processes, it's difficult to get perfect results - only good fibers in the accept flow, only debris in the rejects. Because of this, pulp screening systems are designed in a cascade arrangement (Figure 19). The rejects from the primary screens are processed in the secondary screens to recover good fibers. The rejects from the secondary screens go to tertiary screens for further processing. The rejects from tertiary screening contain very little usable fiber and are removed from the process. Accepted pulp from the secondary screens is fed back to the primaries, and accepts from the tertiaries go back to the secondaries.

After screening, the pulp slurry is at a relatively low consistency, sometimes as low as 1-2%. To minimize storage tank size, the pulp slurry is thickened. The equipment for this, typically called a brownstock decker (Figure 20), simply removes enough water from the pulp to take it up to 10-15% consistency. This medium consistency stock is then stored in a "high-density" storage tank (Figure 21). Brownstock storage separates the unbleached side of the mill from oxygen delignification and bleaching, or from a paper mill making unbleached kraft paper or paperboard.

4.8 SUMMARY

Kraft pulping is a chemical process which converts wood chips into pulp using a solution of sodium hydroxide and sodium sulfide called white liquor. The pulping is done in a batch or continuous digester. Heat is added via direct or indirect steam and delignification occurs. The cooked chips and spent liquor are discharged from the digester to a blow tank, becoming pulp along the way.

In bleachable-grade mills, most of the pulp discharged from a digester is already in the form of individual fibers. If a high-kappa grade is being produced, some mechanical action in a refiner is required to convert any partly-cooked chips into pulp. A succession of separation processes follows: knotters to remove "chunks" of uncooked wood (in bleachable-grade mills), brownstock washing to remove black liquor for chemical recovery, and brownstock screening to remove fiber bundles (also known as shives). The washed, screened pulp is accumulated in a high-density storage tank.

Kraft pulping is controlled to achieve a target kappa number, the standard measure of how much residual lignin is still in the pulp fibers after cooking. The time and temperature of cooking are accounted for in the H-factor; the higher it is, the greater the amount of delignification and the lower the pulp yield. Depending on its intended use, the washed and screened kraft pulp may then be bleached.

4.9 REVIEW QUESTIONS

You will know that you have understood the materials in this chapter if you can easily answer the following questions:

1. What does the kappa number measure?
2. What are the differences between white liquor and black liquor?
3. What are the two main types of digesters?
4. What is the H-factor and why is it useful?
5. What is the purpose of the disc refiner which treats the pulp when it leaves the digester in a high kappa number kraft pulp mill?
6. Why is brownstock washing necessary?
7. Describe two types of pulp washers.
8. Why does brown stock need to be screened?
9. Briefly describe a pressure screen and how it works.
10. Why does the pulp need to be thickened before storage in a high-density storage tower?

Chapter 5

CHEMICAL RECOVERY

OBJECTIVES

At the conclusion of this section, the student will be able to:

- ***Describe the kraft chemical recovery process.***
- ***Explain the purpose of each step in the chemical recovery process, including evaporation, recovery furnace operation, green liquor clarification, dregs removal, slaking/causticizing, white liquor clarification, mud washing, and calcining.***
- ***Describe white liquor, black liquor, green liquor, weak wash, lime mud, reburned lime, and makeup chemicals.***
- ***Describe the 'ABC' test and use its measurements to determine various white/green liquor strength properties.***
- ***Describe environmental controls used in the recovery process.***

5.1 KEY TERMS

ABC test - a lab test for determining the concentration of white or green liquor. The results of the ABC test are expressed in terms of equivalent sodium oxide, Na_2O. The test is a titration of a known volume of a liquor sample with 1.0N hydrochloric acid to a phenolphthalein and methyl orange endpoint. The amount of hydrochloric acid consumed in the titration is used to calculate the liquor concentration, usually in pounds per cubic foot (#/cu.ft.).

Active alkali - sodium hydroxide and sodium sulfide content of white/green liquor.

Activity - the ratio of active alkali to total alkali in white or green liquor, expressed as a percentage.

Alkali- a substance (such as a hydroxide) that neutralizes acid.

Black liquor - the combination of spent cooking liquor and dissolved lignin remaining after cooking chips using the kraft process.

Brownstock washer - a device used to wash black liquor from pulp.

Calcining -converting the lime mud back into lime by heating in the lime kiln.

Causticizer - a stirred retention tank used for the conversion of green liquor to white liquor.

Causticizing - the process of turning green liquor into white liquor.

Caustic soda - NaOH

Cellulose - the main constituent of woody plants and the main component of the fibers used for making paper.

Chemical recovery - a process by which spent pulping chemicals are recovered and reused.

Chemical recovery unit - a furnace in which heavy black liquor is incinerated to recover chemicals and produce steam.

Condensate - steam that has been condensed into hot water.

Direct-contact evaporator - a type of evaporator that brings black liquor in direct contact with steam to concentrate the liquor.

Dissolving tank - an agitated tank in which smelt from a recovery furnace is dissolved with weak wash to make green liquor.

Dregs - bits of carbon and other suspended particles, which are a contaminant in green liquor.

Economizer - a set of tubes on the exit side of the boiler. As the hot gases leave the boiler, some of the remaining thermal energy is transferred via this unit to heat the feed water to the boiler.

Effective alkali - sodium hydroxide content plus one-half of the sodium sulfide content of white or green liquor.

Electrostatic precipitator - a device that removes small particulates from a gas stream by electrical charge.

Evaporators - equipment in the kraft recovery process that increases the solids content of the black liquor from 15-20% to 50-60% so that it can be burned in the recovery boiler.

Extractives - nonfibrous components such as resins and oil which make up 3-10% of wood.

Green liquor - liquor that is formed by dissolving the smelt from a recovery furnace with weak wash. Green liquor contains primarily sodium carbonate and sodium sulfide.

Green liquor clarifier - a device used to remove dregs from green liquor.

Grits - unreacted lime and impurities that are removed in the classifier section of a slaker.

Heavy black liquor - black liquor having a dissolved solids content higher than 65% and capable of sustaining combustion in a recovery furnace.

Hemicelluloses - nonfibrous material, primarily wood sugars, that makes up 15-35% of wood.

Inorganics - refers to the sodium salts in the black liquor, which are reclaimed in the recovery process.

Kraft pulping - an alkaline pulping process that uses sodium sulfide and sodium hydroxide to produce pulp from wood chips.

Lignin - the organic material that holds cellulose fibers together in wood.

Lime - calcium oxide (CaO) produced in alime kiln from lime mud or purchased as a makeup chemical. Lime is used to react with green liquor to produce white liquor and lime mud.

Lime kiln - a device used to convert lime mud into lime.

Lime mud - calcium carbonate that is formed in the causticizing reaction between green liquor and lime.

Makeup chemicals - lime, sulfur, and sodium chemicals added to the liquor cycle in the kraft process to maintain the concentration of the liquor streams.

Multiple-effect evaporators - a series of evaporators (usually five or six) used to concentrate weak black liquor by boiling off water using steam.

Organics - the dissolved chemicals from wood.

Recovery boiler - a boiler that burns black liquor to recover inorganic chemicals used in the cooking process and to produce steam for power generation and for use elsewhere in the process.

Saltcake - common name for sodium sulfate, a chemical used to make up chemical losses in the kraft recovery process.

Slaker - an agitated tank in which lime is mixed with green liquor.

Slaking - the reaction of lime with the water in green liquor to form calcium hydroxide. This is then used to convert sodium carbonate to sodium hydroxide.

Smelt - molten inorganic chemicals that remain after the organic material (lignin) in the black liquor is burned.

Sodium carbonate - Na_2CO_3

Sodium oxide - Na_2O

Sodium sulfide - Na_2S

Strong black liquor - black liquor with a dissolved solids concentration of about 50%.

Sulfidity - the ratio of sodium sulfide to total alkali in white or green liquor, expressed as a percentage.

Tall oil - a by-product of the kraft pulping process which is used to make detergent, soap, and shampoos.

Total alkali - sodium hydroxide, sodium sulfide, and sodium carbonate content of white or green liquor.

Weak black liquor - black liquor with a dissolved solids concentration of about 15%.

Weak wash - a dilute solution of white or green liquor that comes from mud washing and dregs washing operations. Typically the strength is 1/10 or less that of white liquor.

White liquor - a chemical solution of water, sodium sulfide, sodium hydroxide, and sodium carbonate used in the kraft pulping process to convert wood chips into pulp and black liquor.

White liquor clarifier - a device used to separate lime mud from white liquor.

5.2 INTRODUCTION

One of the big advantages of the kraft cooking process is that the chemicals used to dissolve the lignin from the wood can be recovered and recycled. In the process of recovering the chemicals, the dissolved wood components are burned and the heat energy is used to produce steam. In fact, the energy produced in the chemical recovery process and by the burning of bark and other wood residues can produce two-thirds or more of the total energy required for an integrated pulp and paper mill. Trees not only provide the raw materials to make paper, they provide much of the energy to run the whole process. No other major industry can make this claim. Recycling and reuse are not new concepts for the paper industry.

Figure 1: Black Liquor and White Liquor

In kraft pulping, the chemicals in the white liquor dissolve the lignin so that the fibers can be separated. At the end of the cook, the spent cooking liquor, now called black liquor, contains all the dissolved wood components, of which the major one is lignin. The inorganic chemicals used for cooking are also no longer active, that is, they no longer can dissolve the lignin. The purpose of chemical recovery is to recover and reuse the inorganic chemicals, while at the same time using the dissolved organic wood components to produce energy for the mill.

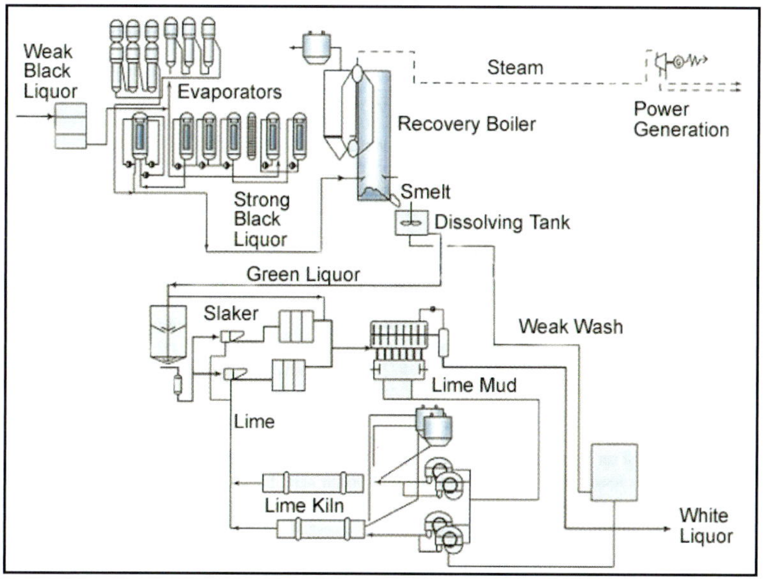

Figure 2: Chemical recovery schematic

In order to recycle the cooking chemicals as white liquor, the black liquor containing the dissolved wood substances and the spent chemicals must be removed from the pulp, and the chemicals converted back to their active state (Figure 1). Black liquor is first washed from the pulp, concentrated in an evaporator, and burned in a recovery boiler. The dried chemicals, which fall to the bottom of the recovery boiler, are called smelt. Smelt is dissolved in water to form green liquor, and the green liquor is converted back into white liquor by treatment with lime. The white liquor is then ready for reuse in the cooking process. Figure 2 shows the chemical recovery process.

Chemical recovery is highly efficient, but some chemicals are always lost in the washing process and the recovery system, so some makeup chemicals must be added. However, since 95% or more of the chemicals are recovered, the cost for makeup chemicals is relatively low. The one important thing to remember about the kraft pulping process is this: without the chemical recovery process and the energy it generates, the kraft process would not be feasible, either from an operating cost or an environmental standpoint.

5.3 BROWNSTOCK WASHING

The function of the brownstock washers is to separate the black liquor from the pulp fibers. Batch digesters are followed by brownstock washers as separate units (Figure 3). Continuous digesters have brownstock diffusion washing inside the digester in addition to separate units after. Rotary vacuum washers are a traditional mainstay of brownstock washing; horizontal belt washers use vacuum to pull liquor through and out of the pulp; diffusion washers use displacement and diffusion to wash liquor from pulp; and pressurized rotary washers use an overpressure developed by air to push liquor through and out of the pulp. Regardless of the type of device, these are collectively known as brownstock washers.

Figure 3: Brownstock washers

The black liquor that is washed from the pulp is collected in filtrate tanks. It is called weak black liquor because it is diluted with wash water and has a solids content of between thirteen and seventeen percent. Black liquor contains two desirable ingredients: fuel and cooking chemicals. Lignin and other wood components dissolved during pulping have a very high heat value, that is, they produce a lot of heat when burned. Figure 4 is a diagram of the brownstock washing process.

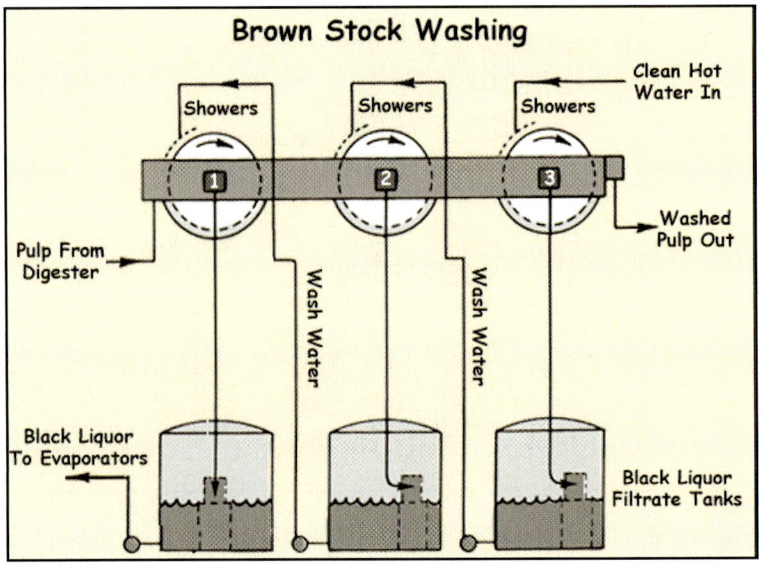

Figure 4: Brownstock washer schematic

5.4 BLACK LIQUOR EVAPORATION

To make the black liquor an efficient fuel, its solids content must be raised to more than sixty-five percent. Most of the required concentration of the black liquor occurs in multiple-effect evaporators. Multiple-effect evaporators are a series of indirect heat exchangers that raise the black liquor temperature to the boiling point in order to remove the excess water as steam. These evaporators are called indirect because the steam and the black liquor are not in contact. Each heat exchanger is called an effect, and normally four to eight effects are arranged in sequence.

Two general types of evaporators are used in the pulp and paper industry: rising-film or falling-film evaporators, which are different kinds of long-tube vertical (LTV) evaporators. LTV evaporators are normally 10 to 15 meters (30 to 50 feet) tall. In a rising-film LTV effect, liquor enters the bottom of the evaporator, rises through the tubes, and is heated by steam surrounding the tubes (Figure 5). The steam transfers its heat to the black liquor through the tube walls, simultaneously condensing into water. This hot water, called condensate, is removed from the bottom of the evaporator shell. As the black liquor rises in the evaporator tubes, it gets hot enough to boil. The steam which is produced from the boiling black liquor is collected in the vapor dome at the top of the unit. This steam produced by the liquor evaporation is used in another evaportor (next effect). The black liquor, which is now at a higher solids content, is transported to the preceding effect. It is important to note that the steam and the black liquor go in opposite directions from one effect to the other. Falling-film effects receive liquor into the top of the effect and allow it to flow as film over the hot plates or tubes. The steam is contained on the inside between the plates, and the black liquor flows downward on the outside of the plates. Falling-film evaporators are more energy-efficient and versatile. Water vapor released is collected at the top of the effect and carried to the next effect. Higher-solids-content liquor is removed from the circulating liquor and sent to the preceding effect.

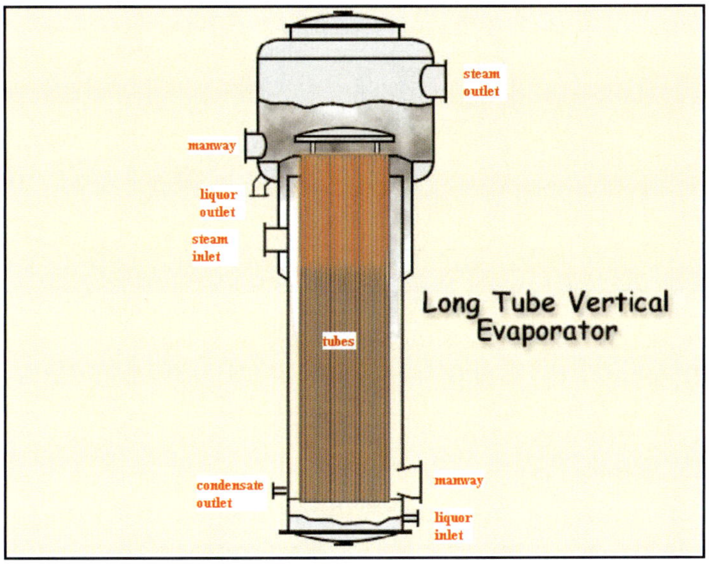

Figure 5: LTV evaporator

Multiple-effect evaporators are a very efficient way to concentrate black liquor, because steam from the boiler is supplied only to the No. 1 effect, which treats the most highly concentrated black liquor, and the steam generated from the evaporation of black liquor in each effect is used in the succeeding effect. The weak black liquor is fed both into the Number 6 and Number 5 effects, then from both of these to the Number 4 effect, and so on. The group of evaporators is numbered based on the direction of steam flow. The steam is supplied to the Number 1 effect, flash steam from there is used in the Number 2 effect, and so on. This countercurrent flow allows the black liquor entering the multiple-effect evaporators at about 15% solids to be concentrated to 50% solids or better. Figure 6 illustrates the steam flow in a multiple-effect evaporator.

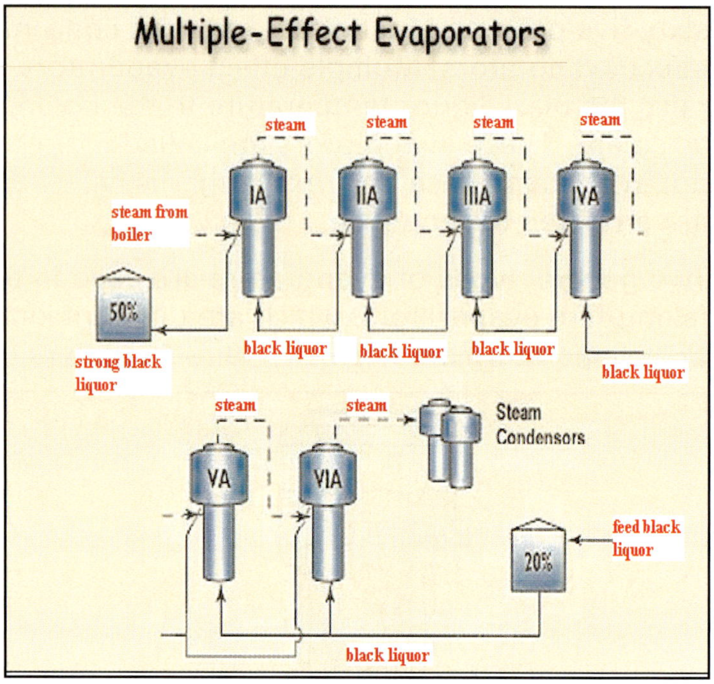

Figure 6: Multiple-effect evaporator schematic

To be burned, the black liquor must be concentrated further, usually to about sixty-five to seventy percent solids. The old way to do this was through direct-contact evaporators. In a direct-contact evaporator, the concentrated black liquor was allowed to contact directly the hot flue gases from the recovery furnace. This method is seldom used now because of the high emissions of sulfur-containing gases from the recovery boiler and the resulting odor. If direct-contact evaporators are used, the black liquor is oxidized using air or oxygen to minimize odor from liquor processing. Most mills now use indirect evaporators called concentrators, which are similar to falling-film evaporators, to remove more water from the black liquor. Concentrators, shown in Figure 7, are more energy-efficient and environmentally friendly than direct-contact evaporators, especially with regard to odor reduction. Sodium sulfate, also known as saltcake, is added to the heavy black liquor prior to combustion in the recovery furnace. Saltcake is a common makeup chemical used to replace the sodium and sulfur lost in the pulping and recovery operations. Other forms of sodium and sulfur are frequently used and may be added at other points in the recovery cycle. As losses from the recovery cycle are reduced, less and less of these makeup chemicals are needed.

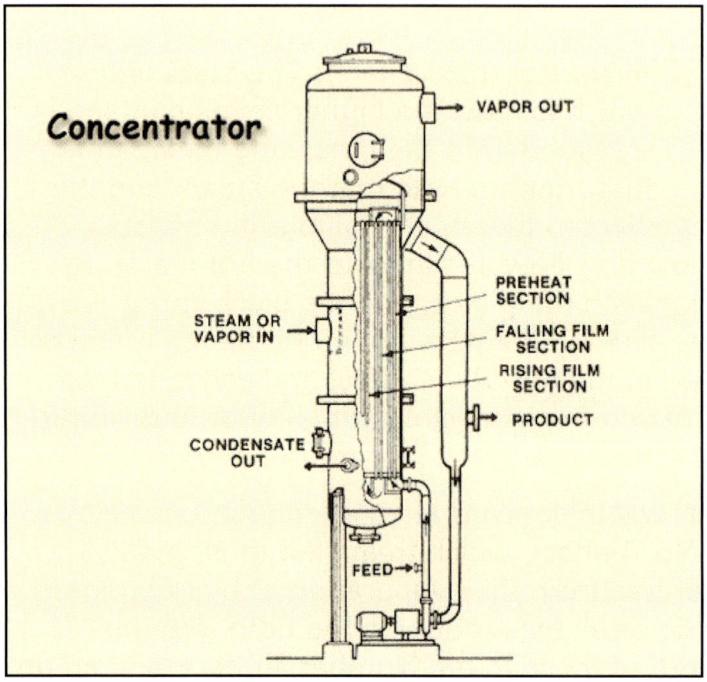

Figure 7: Concentrator

5.5 RECOVERY FURNACE/BOILER

Once the black liquor is fully concentrated, it can be burned in the recovery boiler. The recovery boiler can be thought of as the heart of the recovery process, since it:

- Evaporates the remaining water in the black liquor,
- Burns the lignin and other wood substances,
- Uses the heat of combustion to produce high-pressure steam,
- Begins the conversion of the cooking chemicals back to active form,
- And recovers the cooking chemicals as smelt so that they can be dissolved in weak wash.

Figure 8 shows a typical recovery boiler. The combustion process begins when the concentrated black liquor is sprayed from the liquor guns into the drying zone of the boiler in relatively coarse droplets. Here the rest of the water is driven off the liquor solids, and the dry solids, composed of wood components and cooking chemicals, fall to the bottom of the furnace to form the char bed. Part of the air required for combustion, called primary air, is supplied to the burning char bed, and, as the wood components burn, they convert the sulfur compounds to sulfides. In chemical terms, this is called a reduction reaction, and this lower part of the recovery boiler is therefore called the reduction zone. All other sodium compounds are converted to sodium carbonate. The inorganic chemicals, consisting mostly of sodium carbonate and sodium sulfide, then melt to form smelt, which trickles through the char bed and out of the furnace through a spout into a water-filled tank called the dissolving tank. The molten stream from the recovery boiler looks like lava flowing from a volcano.

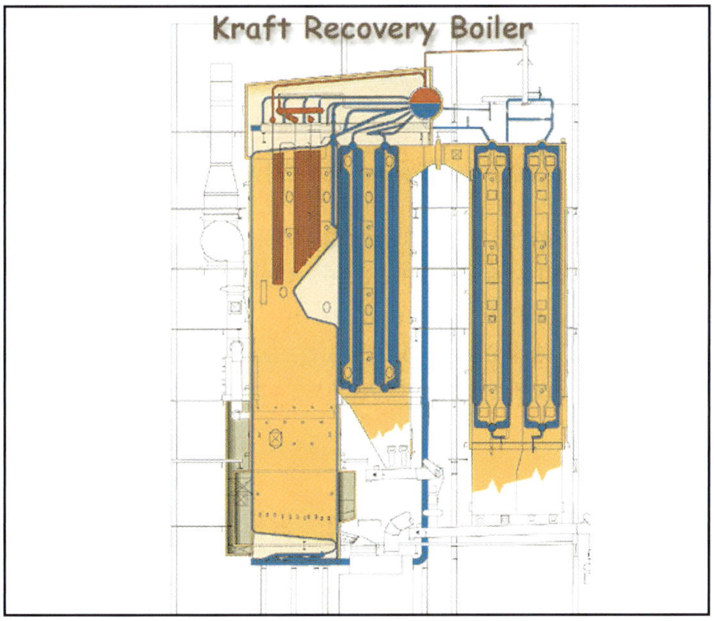

Figure 8: Recovery Boiler

As the black liquor is burned, the temperature of the gases in the boiler can reach 1300 degrees Celsius (2400 degrees Fahrenheit). The walls and floor of the recovery boiler are covered with tubes which contain water or steam. The water in the tubes picks up the heat energy from the burning organic material and turns into steam. In the top part of the boiler, called the superheater, the steam is heated to temperatures well above the boiling point, up to 480 degrees Celsius (900 degrees Fahrenheit).

This superheated steam goes to steam turbines to produce electricity. The steam leaving the turbogenerator, now at reduced pressure, is used elsewhere in the mill - in the evaporators and digesters, to dry pulp or paper, and for various other uses around the mill. Pulp and paper mills, which use a substantial amount of electricity, usually generate over half of their electrical power needs from the steam generated by the burning of black liquor.

As the hot gases rise to the top of the boiler, secondary and tertiary air is added to make sure that the combustion products are fully oxidized. This is another chemical term which means that the gases have been fully burned or fully combined with oxygen. Recovery boilers are quite tall, so that there is plenty of time for the gases to transfer their heat to the water in the boiler tubes and thus cool before exiting the boiler. The gases pass through the superheater (Figure 9), cooling to around 450 degrees Celsius (850 degrees Fahrenheit). There is still a lot of heat energy in the gases that can be reclaimed by the economizers. In the economizer, the incoming boiler feedwater, which is relatively cool, picks up much of the remaining heat from the exhaust gases.

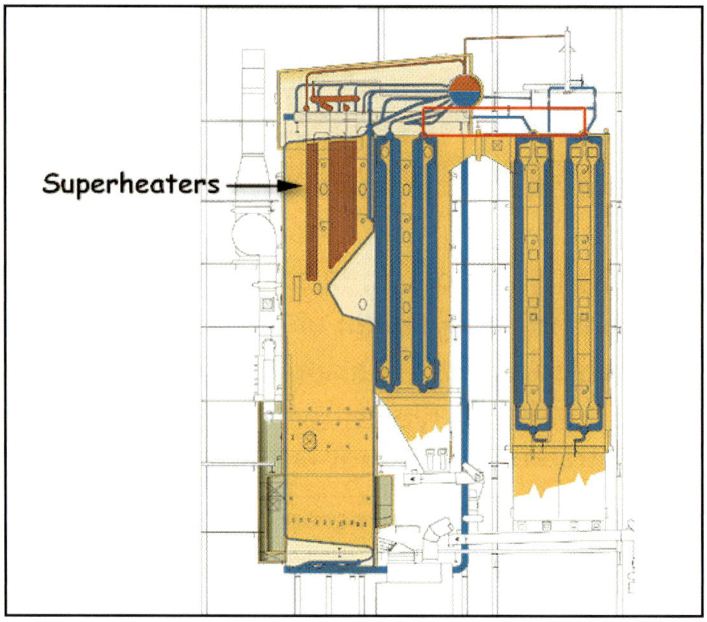

Figure 9: Recovery Boiler superheater

Once most of the heat in the exhaust gases has been captured, they can be released to the atmosphere, but not before they are cleaned of dust particles. The gases pass through the electrostatic precipitators (Figure 10) which remove any particulate matter. These particulates, which contain chemicals, are returned to the heavy black liquor that is being fired into the recovery furnace. The clean gases, which are mostly water vapor and carbon dioxide, can now be released to the atmosphere.

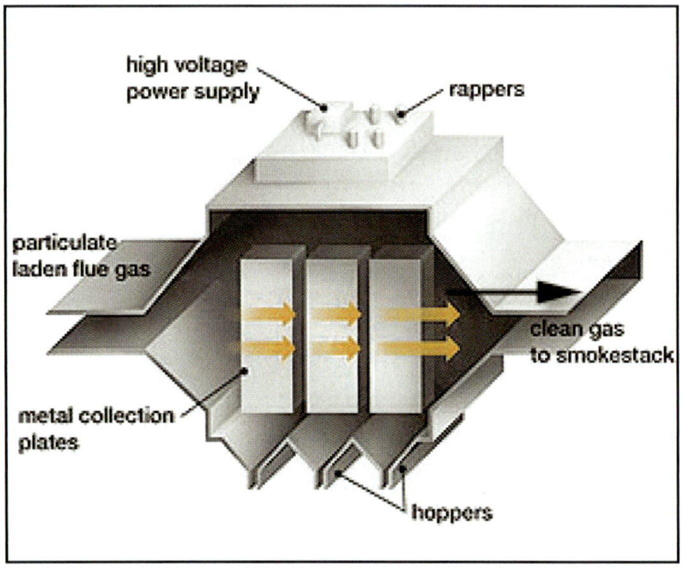

Figure 10: Electrostatic precipitator schematic

5.6 CHEMICAL RECOVERY

The molten smelt flows into the dissolving tank where it is mixed with "weak wash" to form green liquor, which appropriately enough is green in color. The green liquor is contaminated with bits of carbon and other undissolved material. These suspended solids called "dregs" are removed in the green liquor clarifier (Figure 11), where the dregs settle to the bottom and the clarified green liquor is removed from the top. Dregs are then washed and the wash water used in the dissolving tank. The dregs are then disposed of as landfill. Some mills use pressure filters in place of clarifiers.

The recovery boiler started the process of reactivating the cooking chemicals, but one more chemical reaction is needed before the cooking chemicals are ready to be used again to produce pulp. The green liquor is mixed with lime to produce a causticizing reaction. Here the "quick" lime is converted to calcium carbonate (called lime mud) and the green liquor is turned into white liquor. This reaction begins when lime is mixed with green liquor in the slaker. The first step is the slaking reaction, where quick lime reacts with water in green liquor to form milk of lime (calcium hydroxide). Unreacted lime particles called "grits" are raked from the slaker, washed, and disposed of as landfill.

Slaking reaction:

CaO	+	H_2O	$\longrightarrow$	$Ca(OH)_2$
lime	+	water from green liquor		milk of lime

The chemical reaction continues in a series of reactors called causticizers (usually three or four), where the mixture of green liquor and lime is agitated for several hours to form sodium hydroxide and calcium carbonate. After the causticizers, the mixture contains white liquor and lime mud. The white liquor and lime mud are separated in the white liquor clarifier, where the mud settles to the bottom. The white liquor is now ready for the digester.

Causticizing reaction:

Na_2CO_3 +	$Ca(OH)_2$	$\longrightarrow$	$2NaOH$	+	$CaCO_3$
sodium carbonate	milk of lime		caustic		lime mud

In the causticizing process, the lime has been chemically changed, so the lime mud must go through its own chemical recovery process so that it can be used again. First the lime mud is washed to remove any alkali, then washed and thickened on a drum filter to 60% to 70% solids. The lime mud is then dried and burned in the lime kiln to form "quick" lime.

A lime kiln looks like a long rotating horizontal pipe (Figure 12). Typical size is around 2.5 to 4 meters (8 to 13 feet) in diameter and 30 to 120 meters (100 to 400 feet) long. The lime kiln has a slight incline and rotates at about one rpm in order to move the lime slowly from the high or cool end of the kiln to the low or hot end. The low or discharge end of the kiln has a gas or fuel oil burner that supplies the heat energy needed to dry the lime mud and convert it to active or quick lime. The temperature in the combustion end of the kiln can reach 1200 degrees Celsius (2200 degrees Fahrenheit). The exhaust gases from the kiln go through a scrubber or electrostatic precipitator to remove particulate matter before they are released to the atmosphere.

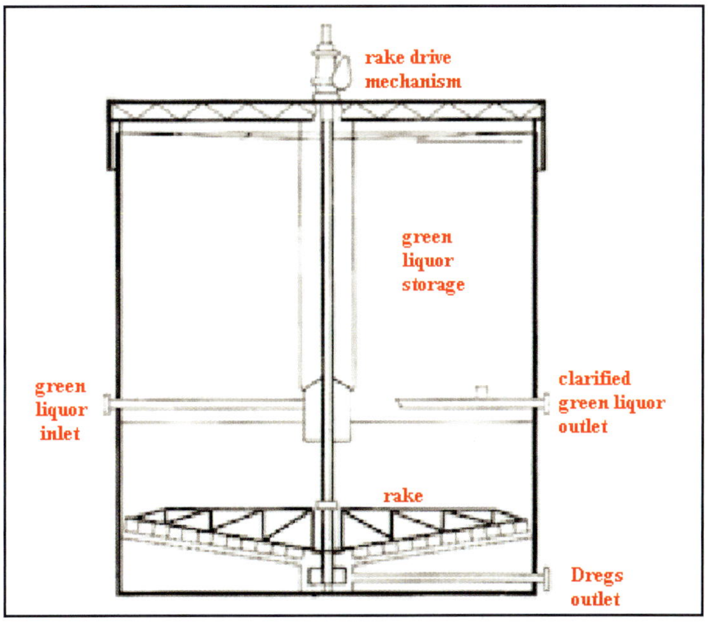

Figure 11: Green liquor clarifier

Figure 12: Lime kiln

5.7 BY-PRODUCTS

One final aspect of chemical recovery is the by-products of the pulping and recovery process. The biggest by-product of pulping is of course the lignin, which is part of the black liquor. Lignin has a very high heat value, and it therefore can supply all the thermal energy needed to recover the chemicals and to provide high-pressure steam for power generation and other steam requirements around the mill. Without lignin, the recovery process would not be economically feasible. It is interesting to note that some mills sell a small amount of lignin for other uses.

Wood species that have a lot of resins or "extractives" can provide other useful by-products. Softwood, particularly Southern pine, contains a lot of turpentine that is driven from the chips in the early stages of the heating process in the digester. In the United States, the paper industry produces around 30 million gallons per year of turpentine. Raw turpentine can be distilled and used in a number of products such as solvents, perfumes, and pressure-sensitive adhesives.

Tall oil is another by-product of the kraft pulping process. Tall is the Swedish word for pine, and tall oil is collected when wood from pine trees is chemically pulped by an alkaline pulping process. Tall oil is collected as a "soap" that is skimmed from the top of the black liquor after the liquor has been evaporated to 25% to 30% solids. After it is purified, tall oil has many uses, including as an ingredient in detergents and shampoos and as a sizing agent in paper.

5.8 CONCLUSIONS

The chemical recovery process is economically vital to the pulp mill for recovery of chemicals, generation of power, and protection of the environment. The process recovers the chemicals used in pulp cooking, may recover wood extractives such as tall oil, and uses the organic material dissolved from wood in the pulping process to generate energy.

The recovery process begins with the collection of the black liquor after it is washed out of the unbleached pulp. The black liquor is evaporated and the resulting heavy black liquor is sprayed into the recovery boiler as fuel. The inorganic chemicals, largely sodium and sulfur compounds, form smelt which is collected from the furnace and sent to a smelt dissolving tank to make green liquor. From the dissolving tank, the green liquor is sent to a clarifier for dregs removal. At the slaker, lime is added and the causticizing process started, which converts the green liquor to white liquor. From there, the white liquor is clarified and returned to the white liquor storage tank for reuse in the digester.

The heat generated by the recovery boiler furnace is used to generate high-pressure steam. The steam is fed into the steam turbine where it is converted to electrical energy and process steam. This production of electrical energy and process steam together is called cogeneration. The lime mud from the white liquor clarifier is returned to the lime kiln for calcining. The resulting lime is then sent back to the recausticizing plant to convert green liquor into white liquor. The gases from the recovery furnace and lime kiln go through an electrostatic precipitator and/or a scrubber to remove particles, and the now-harmless gases, which are primarily water vapor, are released from the mill.

5.9 LIQUOR TESTING

The ABC test is a means of determining the chemical concentration of various mill liquors. The test is a titration using a known concentration of hydrochloric acid (HCl) with phenolphthalein and methyl orange indicators. The concentrations of the liquor are commonly expressed as equivalent sodium oxide (Na_2O), even though there is no sodium oxide present in the liquor.

The 'A' test = effective alkali (EA) = NaOH + ½ Na_2S

The 'B' test = active alkali (AA) = NaOH + Na_2S

The 'C' test = total titratable alkali (TTA) = Na_2CO_3 + NaOH + Na_2S

Various liquor properties related to these concentrations can be determined by manipulating the ABC results:

% sulfidity = % ratio of sodium sulfide to TTA = [2(B - A)/B]100

% activity = % ratio of active alkali to TTA = (B/C)100

Na_2S = 2(B-A)

NaOH = B-2(B-A)

Na_2CO_3 = C-B

*****remember these are reported as equivalent Na_2O*******

Typical mill white liquor concentrations:

Chemical	#/cu.ft. (as Na_2O)
NaOH	5 - 7.5
Na_2S	1.9 - 2.5
Na_2CO_3	0.7 - 2.7

Example of how to calculate the concentrations from the ABC test:

An operator has just run the test on a sample of white liquor with the following results:

A = 14.2 ml of 1.0N HCl

B = 16.8 ml of 1.0N HCl

C = 20 ml of 1.0N HCl

When using 1.0N HCl, the factor to convert ml of 1.0N HCl to #/cu.ft. of Na_2O is 0.387.

TTA=C=20= 20(0.387) = 7.7 #/cu.ft

AA =B=16.8= 16.8(0.387) = 6.5 #/cu.ft.

EA =A=14.2= 14.2(0.387) = 5.5 #/cu.ft.

% sulfidity = (2(16.8-14.2)/20)100 = 26.0%

% activity = (16.8/20)100 = 84.0%

Na_2S = 2(16.8-14.2)(0.387) = 2.0 #/cu.ft.

NaOH = (16.8-2(16.8-14.2))(0.387) = 4.5 #/cu.ft.

Na_2CO_3 = (20-16.8)(0.387) = 1.2

5.10 REVIEW QUESTIONS

1. Calculate the % sulfidity and % activity of green liquor using the following ABC test results:
   A = 5.0
   B = 7.5
   C = 20.5
2. Name different types of brownstock washers.
3. State the objective of brownstock washing.
4. List the main steps of the recovery process in the kraft recovery furnace.
5. List the contents of a typical kraft mill black liquor.
6. What is the objective of black liquor evaporation?
7. Describe the operation of a multiple-effect evaporator system with respect to liquor and steam flow.
8. Draw the liquor cycle and name the primary operations and liquors.
9. What happens in the causticizing process?
10. Draw and identify a box diagram of a typical causticizing system, using arrows to show the flow paths.
11. How is green liquor formed?
12. Write the slaking and causticizing reactions.
13. What is the purpose of green liquor clarification?
14. What is the purpose of white liquor clarification?
15. Describe the operation of a lime kiln.

16. What are dregs?

17. Match the following liquor properties with the appropriate definition of each:

a.	A	1. active alkali
b.	B	2. total alkali
c.	C	3. effective alkali
d.	Na_2S	4. caustic
e.	Na_2CO_3	5. sulfide
f.	NaOH	6. sodium carbonate
g.	grits	7. girls raised in the south
h.	dregs	8. unreacted lime
i.	white liquor	9. cooking liquor
j.	green liquor	10. smelt dissolved in weak wash
		11. carbonaceous impurities in green liquor
		12. sodium oxide

Chapter 6

MECHANICAL AND CHEMI-MECHANICAL PULPING

OBJECTIVES

At the conclusion of this section, the student will be able to:

- ***Explain the differences between mechanical and chemical pulping processes.***
- ***Describe the mechanical and Chemi-mechanical processes.***
- ***Describe the differences between mechanically-produced and chemically-produced pulps.***
- ***List the advantages and disadvantages of mechanical pulping processes compared with chemical pulping processes.***

6.1 KEY TERMS

APMP - Short for alkaline peroxide mechanical pulping, a Chemi-mechanical process that uses hydrogen peroxide in the initial chemical treatment to soften the lignin and brighten the pulp.

BCTMP - Short for bleached chemithermomechanical pulping, a CTMP pulp that has also been bleached to improve brightness.

Brightness - Generally refers to the whiteness of the pulp or paper.

Chemical Pulping - A process to dissolve lignin by the use of chemicals in order to liberate the cellulose fibers from wood.

Chemi-mechanical Pulping - A process that combines a mild chemical treatment of chips with a mechanical one to liberate the fibers.

Consistency -The percentage by weight of ovendry fibrous material in a stock or stock suspension, sometimes called density, solids, or concentration.

Corrugating Medium -The fluted paper between the two layers of linerboard in a corrugated box.

CTMP - Short for chemithermomechanical pulping, a hybrid high-yield pulping process that uses a mild chemical treatment followed by high-temperature mechanical action to separate cellulose fibers from wood.

Groundwood - A high-yield mechanical pulp produced by pressing logs sideways against a rotating pulp stone.

Hydrogen Peroxide - A chemical used to bleach pulp, often used as a pre- or post-treatment for mechanical pulps.

Impregnator - a piece of equipment that helps force pulping chemicals into wood chips.

Kraft Pulping - an alkaline pulping process that uses sodium hydroxide and sodium sulfide to dissolve lignin and release cellulose fibers.

Latency Chest - A tank or chest that gently agitates mechanical pulps in order to allow the twists and kinks in the fibers to dissipate.

Lignin - A substance that acts as a glue between cellulose fibers to form the woody structure of plants.

Mechanical Pulping - A high-yield pulping process which depends primarily on mechanical action to separate fibers with minimal removal of lignin.

Multi-Ply Paperboard - heavyweight grades of paper made up of two or more layers of different types of pulp, used in milk cartons, cereal boxes, paper plates and cups, shoeboxes, etc.

NSSC - Short for neutral sulfite Semi-chemical, a high-yield pulping process that combines chemical treatment with mechanical defibering; used to make pulp for corrugating medium.

Opacity -The ability of paper to obstruct the passage of light, which prevents printing ink on one side of the paper from being seen from the other side.

PGW - Short for pressurized groundwood, a stone groundwood produced under pressure to improve fiber strength.

Pulping -The process of separating a cellulose-containing raw material, such as wood or straw, into individual fibers which can be reformed to make paper.

Refiner - Rotating equipment with discs having bars on their surfaces that, in mechanical pulping processes, separate fibers from wood chips.

RMP - Short for refiner-mechanical pulp, the first mechanical process using refiners to separate fibers from wood chips.

SGW - Short for stone groundwood (see groundwood).

Shives - Small splinters of wood that may be present in pulp as contaminants.

TMP - Short for thermomechanical pulping, a process that combines heat and mechanical action using a disc refiner to liberate fibers from wood chips.

Yield -The percentage of usable product produced from the raw material used in any process.

6.2 INTRODUCTION TO MECHANICAL AND CHEMI-MECHANICAL PULPING PROCESSES

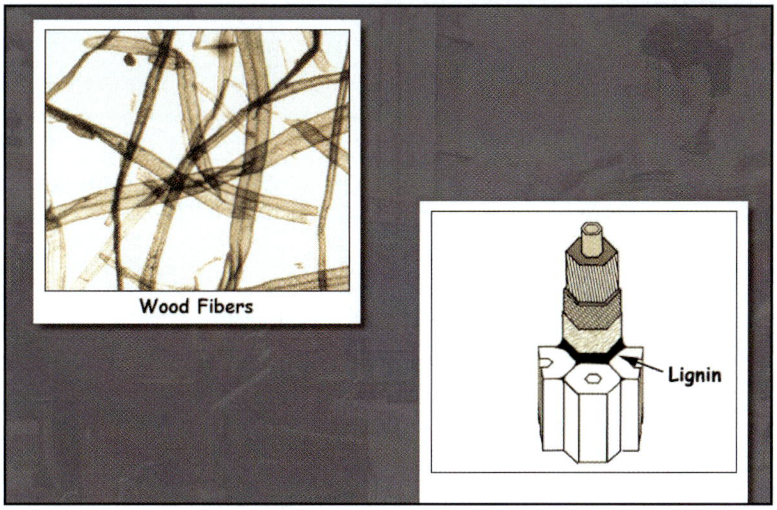

Figure 1: Wood fibers and lignin

Paper is made from wood fibers, and wood fibers are held together by a naturally-occurring glue in wood, called lignin. (Figure 1) There are two ways to separate the fibers so they can be used to make paper: one is to separate fibers from the wood mechanically, the other is to dissolve the lignin chemically. This second option will be described in the lesson on chemical pulping. As you might expect, both mechanical and chemical processes can be added together to form hybrid or Chemi-mechanical pulping.

For the first two thousand years or so of papermaking, vegetable fibers were reused to make paper. Cotton and linen rags, old fishing nets, and discarded hemp rope were some of the fiber sources for early papermakers. But you can see how fiber supply could start to become a problem. Trees represent a good source of fiber, but how do you get the fibers apart? The answer to that question was something called stone groundwood. The stone groundwood pulping process was invented over 150 years ago. It was the first pulping process that separated wood fibers for use in papermaking, and it was the birth of mechanical pulping. (Figure 2)

Figure 2: Stone groundwood

While kraft chemical pulping now dominates the pulp and paper industry, mechanical pulping and Chemi-mechanical processes are increasingly popular. Why? Because there are inherent advantages to mechanical and hybrid pulping, and the big one is yield. Yield refers to how much pulp is produced from a given amount of wood. For

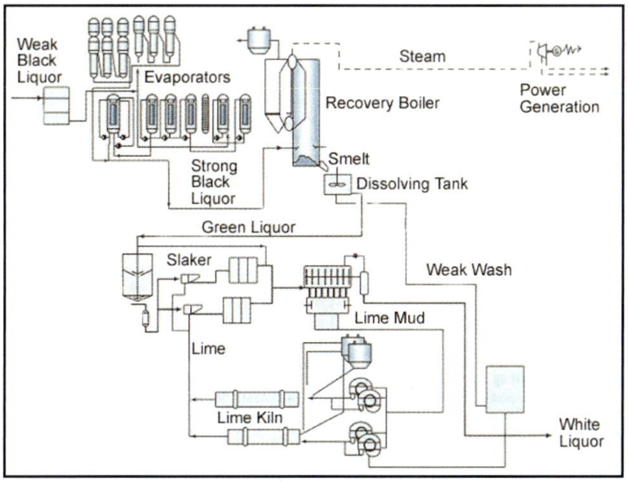

Figure 3: The pulping process

example, if 1000 pounds of wood produces 500 pounds of pulp, then the yield would be 50%. Typical yield for a kraft pulp mill is below 50%, but yield for mechanical and hybrid pulps may be between 75 and 95%, and high yield means lower costs.

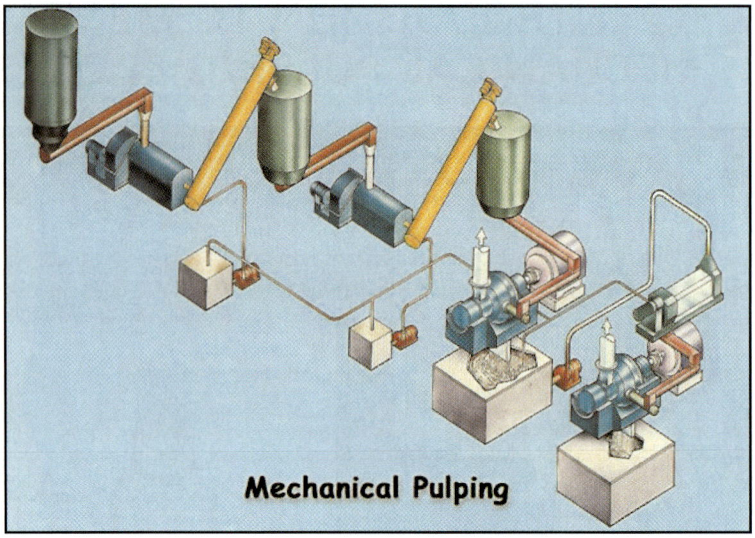

Figure 4: Mechanical pulping flow diagram

Another advantage is the relative simplicity of mechanical and Chemi-mechanical pulping processes compared to the kraft process. (Figure 3) Kraft pulping requires a chemical recovery plant, which is both complex and expensive. Kraft mills must be large, that is, produce at least 1000 tons per day, to be economically feasible. Mechanical pulping units can be sized much smaller to fit the needs of the paper mill and still make economic sense. Figure 4 provides a schematic of the mechanical pulping process.

Since mechanical pulping does not require a chemical recovery process, a mechanical pulp mill does not have to contend with the complex air-related environmental issues that are critical for a kraft mill. So capital costs can be much lower for a mechanical or Chemi-mechanical pulp mill. On the other hand, operating costs for a mechanical pulp mill can be high because the process does not produce its own energy as the kraft recovery process does. Typically, electrical costs are a large part of the cost of making mechanical pulp, so mechanical pulping makes better economic sense in locations where electrical rates are low.

Figure 5: Newspapers

Finally, there are advantages as well as disadvantages to the type of fiber that is produced in mechanical and Chemi-mechanical pulping. High yield means that much of the lignin remains with the pulp, so mechanical pulps do not bond as well as chemical pulps and therefore produce weaker paper. On the other hand, mechanical pulp fibers are bulky, stiff, and have high opacity, all of which can be good properties for many grades of paper. One inherent disadvantage is that the remaining lignin prevents the fibers from being bleached to the high brightness that kraft pulps can attain. Mechanical pulps can be brightened by chemical treatments such as sodium sulfite, sodium hydrosulfite, or hydrogen peroxide, but the brightness will revert to lower values when the paper is exposed to sunlight and air. Think of what happens to your newspaper when it is left in the sun for a day or so. (Figure 5)

Mechanical pulps are best suited for grades of paper that do not require permanence, such as newsprint or magazine paper, for use in multi-ply paperboard (Figure 6) as a center ply where it can be covered by chemical pulps, or in a mixed furnish with chemical pulps.

6.3 MECHANICAL PULPING: THE STONE GROUNDWOOD PULPING PROCESSES

Stone groundwood or SGW has been used for newsprint and magazine papers for many years, and is usually mixed with chemical pulp to provide the strength needed in the final paper. The discovery of grinding wood as a method to make paper occurred in 1838 - 1840. Prior to that, paper was made only from cooking and bleaching old rags.

Pressing the sides of short logs against a rotating abrasive wheel produces SGW (see Figure 2). Remember that fibers are shaped like drinking straws with their length aligned with the long dimension of the logs, so the fibers are lying against the face of the stone. Fibers are held together by lignin, and the friction of the groove pattern on the stone produces heat, which causes the lignin to soften. Fibers are not actually ground off the log, as you might think. Instead, the softening of the lignin as well as the compression and decompression of the area around the fibers causes them to separate from the log.

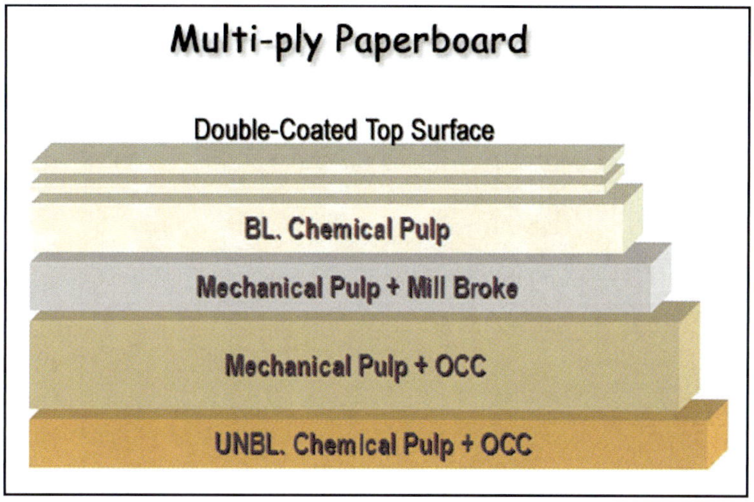

Figure 6: Multi-ply paperboard

The SGW process generally results in pulp with very high opacity but also the shortest fibers of any pulp. Because of these characteristics, SGW pulp produces paper with excellent printing properties but low strength. The SGW process works better on softwoods since the wood is softer and can be more easily compressed. Spruce and fir, which grow in colder climates, are ideal for SGW because of their light-colored wood and their low level of resins.

As you might expect, there are a lot of different configurations for grinders. A rotating grinding stone is located in the grinding pit. Logs, which can be five or six feet long, are fed through a magazine into a hydraulically-loaded grinding pocket. Grinding pressure and temperature are key operating variables. During the grinding process, the temperature of the stone must be kept under control or the wood will be charred. The operator controls the grinding temperature with cooling showers. Cooling showers wash the fibers from the stone while maintaining the process temperature at 140 to 180 degrees Fahrenheit. Sometimes the stone is partially submerged in the water in the grinding pit to help control the temperature. Depending on the amount of shower water, the consistency (that is, the

Figure 7: Pressure screens

Figure 8: Centrifugal cleaners

percent solids) of the fibers in the pit is 2% to 4 %. As fibers exit the pit, they flow over a dam into a storage chest. After the storage chest, the pulp is screened to remove unseparated fiber bundles called shives. Either in the pulp mill or prior to the paper machine, SGW pulp is screened and cleaned to remove dirt and grinder grit from the finished product using pressure screens (Figure 7) and centrifugal cleaners (Figure 8).

After thickening, the fibers may be brightened with hydrosulfite or peroxide and then stored for use on the paper machine. The stones, which used to be sandstone, are now made up of segments of aluminum oxide or silicon carbide attached to a concrete core. Stones become dull after several days and have to be resharpened with a burring wheel. Immediately after the stone is sharpened, pulp strength decreases until the stone is "broken in." However, since a mill typically has a number of grinders, sharpenings can be rotated among the grinders to keep the overall pulp strength consistent.

Although the SGW process is largely unchanged since its invention over 150 years ago, there has been one significant modification. If the grinding process is pressurized with steam at 105 to 120 degrees Celsius (220 to 250 degrees Fahrenheit), the fibers separate more easily. As the logs enter the first chamber, the door closes to maintain pressure when the second chamber is opened and the logs fall against the rotating stone. Pressurized groundwood, or PGW, produces a higher-quality pulp at lower energy consumption compared to SGW.

6.4 MECHANICAL PULPING: REFINER-MECHANICAL PULPING PROCESSES

One of the difficulties with SGW is that whole logs must be pressed against the grinding stone. However, wood has more economic value if the logs are first cut into lumber and the residual wood is then processed into chips for paper mills. Another problem with the SGW process is that the pulp produced has low strength and high fines content. Fines are small pieces of fiber that have been broken off. High fines content means the pulp has excellent opacity but is difficult to retain on the wire and drains slowly in the forming process on the

Figure 9: Refiners

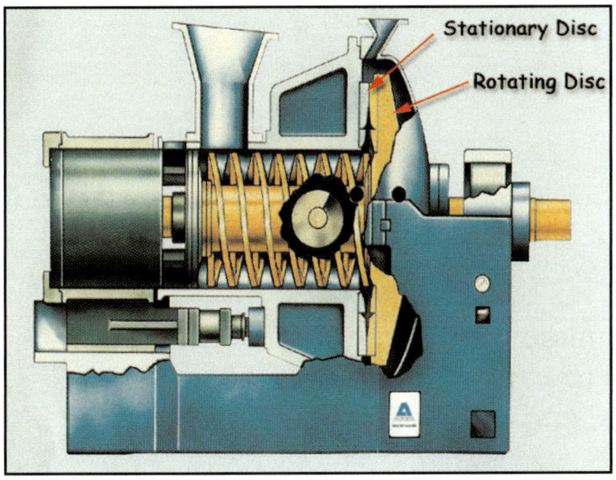

Figure 10: Cross-section of disc refiner

Figure 11: Disc refiner plates

papermachine. Both of these problems were addressed in the 1930s with the refiner-mechanical process, or RMP. RMP used chips instead of logs and disc refiners in place of a grinding stone. In a later development, steam was added in the first-stage refiners. This process became known as thermomechanical pulping, or TMP.

The refiners are the essential part of any mechanical pulping process (Figure 9). A refiner consists of one or more rotating discs (Figure 10). Each disc has a surface covered with refiner plates (Figure 11) that have raised bars. In a single-disc refiner, one set of plates is fixed, that is, it does not rotate, and the second set of plates is attached to a rotating disc. These are sometimes called the stator and the rotor. Refiners may also be configured with twin refining zones on either side of the rotor. Whatever the configuration used, the plates are set close together, and the wood chips are separated into individual fibers as they pass between the moving bars. Plate clearance obviously is very important, and this distance between the plates is controlled by electromechanical or hydraulic motors. There is a large opening in the center of the refiner that allows chips in and pulp out. Chips are fed in using a ribbon feeder which feeds the chips into the center of the refining plates.

Wood chips have one disadvantage compared with logs, and that is the amount of dirt and debris that chips contain. So chip washing is a necessary first step. The process shown in Figure 12 uses a rotary pulp washer. Chips are drained in a screw drainer and then fed to a transfer conveyor leading to a storage bin. The chips are then fed through a plug screw feeder and conveyed to the pressurized refiner. Water that is discharged from the screw drainer is sent to a sand separator to remove sand and heavy particles, then to an inclined separator to remove fines so the water can be reused in the chip washer.

Lignin, the glue that holds the fibers together, is similar to hot-melt glue, since it softens when you heat it. Steam is therefore added to raise the temperature of the chips to about 90 degrees

Celsius (195 degrees Fahrenheit) (Figure 13).

In the refining process, the friction generates additional steam and pressure. Normally, a refining temperature of 120 to 130 degrees Celsius (250 to 265 degrees Fahrenheit) is sufficient to soften the lignin so that the fibers can be separated intact. Wood species is also important. Due to fiber length and better strength of the product, softwood is preferred for the TMP process.

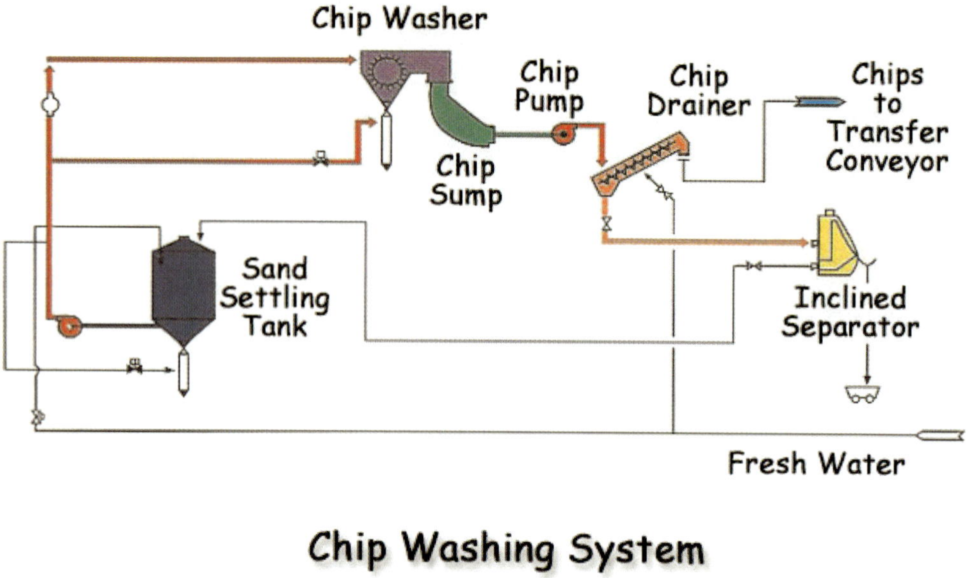

Figure 12: Chip washing system

Chips are broken up and the fibers are separated as they pass between the rotating bars on the refiner plates. If you look at the plates in Figure 14, you can see a section near the inside of the disc with large, widely spaced bars.

Since this is where the chips enter the refiner, these coarse bars begin the disintegration of the chips. As the broken pieces of chips move further towards the top or outside of the plate, they have to pass through an intermediate zone, and finally through a section with finely spaced bars.

In the process shown in Figure 15, the chips pass through a pressurized primary refiner that starts the disintegration of the chips. The pulp then passes through a cyclone separator to

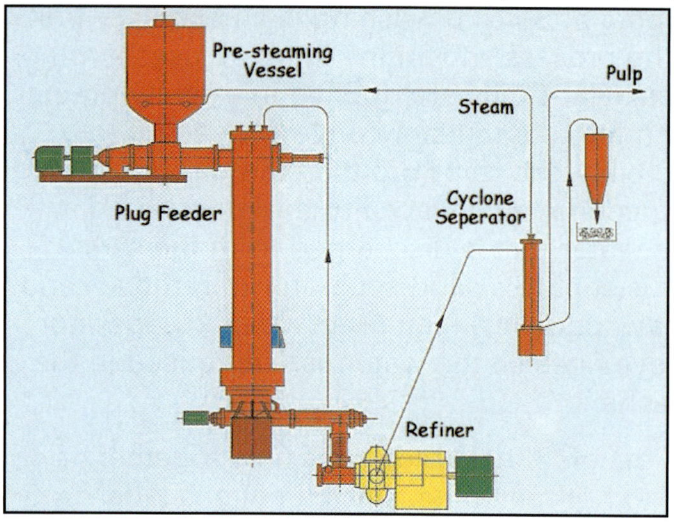

Figure 13: Steam added

Figure 14: Disc refiner plate patterns

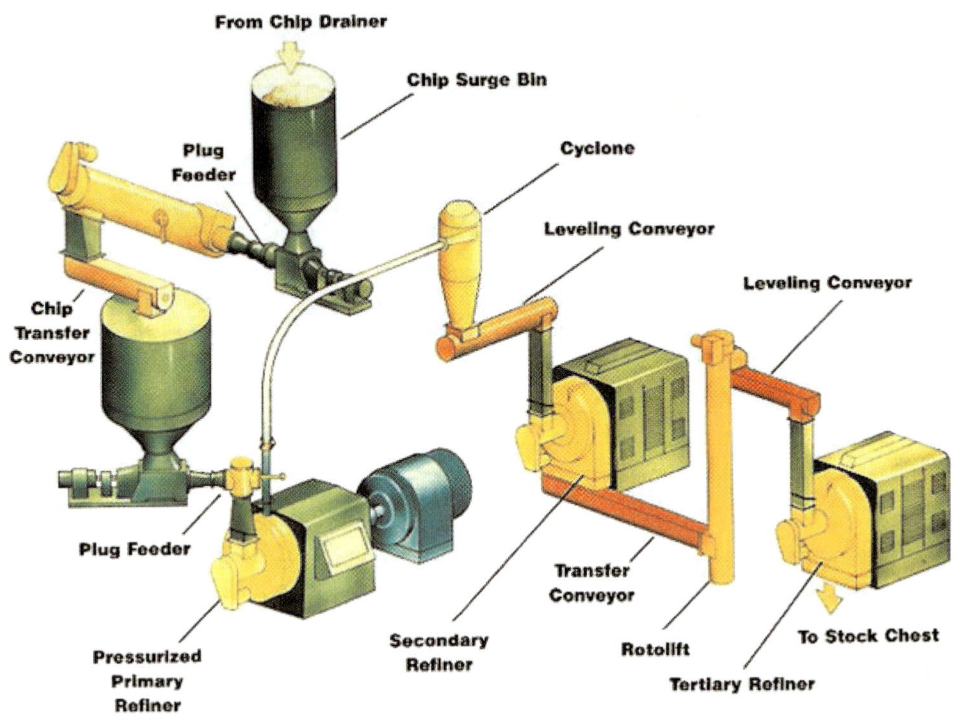

Figure 15: Pressurized primary refiner

allow the steam generated in the refining process to be reused. Some mills also have an inter-storage washing stage with a pressing step to help remove pitch and resin and dissolved solids. Then the pulp enters the pressurized secondary refiners, which are similar to the primary refiners except for the refiner plates, which have more closely spaced bars. By the time the chips have passed through the secondary refiner, they are completely separated into individual fibers. Well, "completely" isn't exactly the right word, because there are always fiber bundles or shives that are too large to make into paper. Shives must be separated from the good fibers, and this is done in several stages of pressure screens. A pressure screen is shown in Figure 16.

The screen basket (Figure 17) has holes or slots that the good fiber can pass through. A rotor on the inside of the basket has blades that create a high-frequency negative pressure pulse that keeps the screen from plugging. Rejects from the screening operation are sent to the reject refiners, then screened and cleaned and returned to the primary fiber line.

Just to give you some sense of size, in the process just described, the refiners have 1600-millimeter (63-inch) diameter refiner plates, and each refiner might be connected to a 25,000-horsepower (19-megawatt) motor. The largest refiners in industry today can accommodate a single 47,000-horsepower (35-megawatt) motor. That is a lot of energy! Much of this energy is transformed into heat energy in the refiners. In order to recover the heat from the process, the steam that is generated is separated from the pulp in a pressurized cyclone. Some of this steam might be used directly, otherwise it is sent to a reboiler that makes clean steam from the

Figure 16: Pressure screen

Figure 17: Screen basket

heat in the contaminated steam. Many TMP plants provide up to 75% of the steam needs of the pulp mill and paper mill from the steam generated in the refining operation. Even with steam recovery, the TMP process consumes a lot of energy in the form of purchased electrical power, so the price of purchased power is really important to the economic success of a TMP mill.

6.5 CHEMI-MECHANICAL AND SEMI-CHEMICAL PULPING

So far, we have addressed purely mechanical separation of fibers. In the chapter on kraft pulping, we talk about chemical separation of fibers. But why not a little of both? Heating of wood results in a temporary softening of the lignin, but chemical treatment can result in more permanent softening of the lignin. Pulping processes that add a chemical treatment to the mechanical one are referred to as Chemi-mechanical or Semi-chemical pulping, depending on how strong the chemical treatment is. These pulping processes have many advantages. In fact, combination processes are the fastest-growing type of pulping. As previously indicated, thermomechanical pulp or TMP has replaced a lot of the pulp that used to be produced using the SGW process. Wood is more valuable if it is first made into lumber and the residuals turned into chips for the pulp mill. TMP is an improvement over SGW.

By adding a chemical treatment to the TMP process, we get something called chemithermomechanical pulping or CTMP. CTMP is a common pulping process because it results in a higher-quality fiber with better brightness than purely mechanical pulping processes. A real advantage is that hardwoods as well as softwoods can be pulped using CTMP, and the pulp produced can be more easily bleached to brightness levels close to high-brightness kraft pulps. Chemical treatment removes a lot of the resins and some of the hemicelluloses, so yield drops to the 85% to 90% range, but the pulp that is produced is much stronger than straight TMP and can be used in place of higher-cost kraft pulp. If you look at a CTMP process flowsheet (Figure 18), you will see that the process is similar to TMP except for the chemical pretreatment stage.

In the process shown, the chips are presteamed in the chip surge bin and then compressed in a plug screw feeder. Sodium sulfite solution is applied as the chips are released to the impregnator. In the impregnator, the chemical retention time varies from one minute to five minutes. Then the chips are drained and discharged into the reaction bin. Chips are then broken into fibers in the following stages: first a pressurized primary refiner, then a pressurized secondary refiner, and finally an atmospheric-pressure tertiary refiner. The cyclone after the primary refiner is used to separate the steam from the fibers so the steam can be reused.

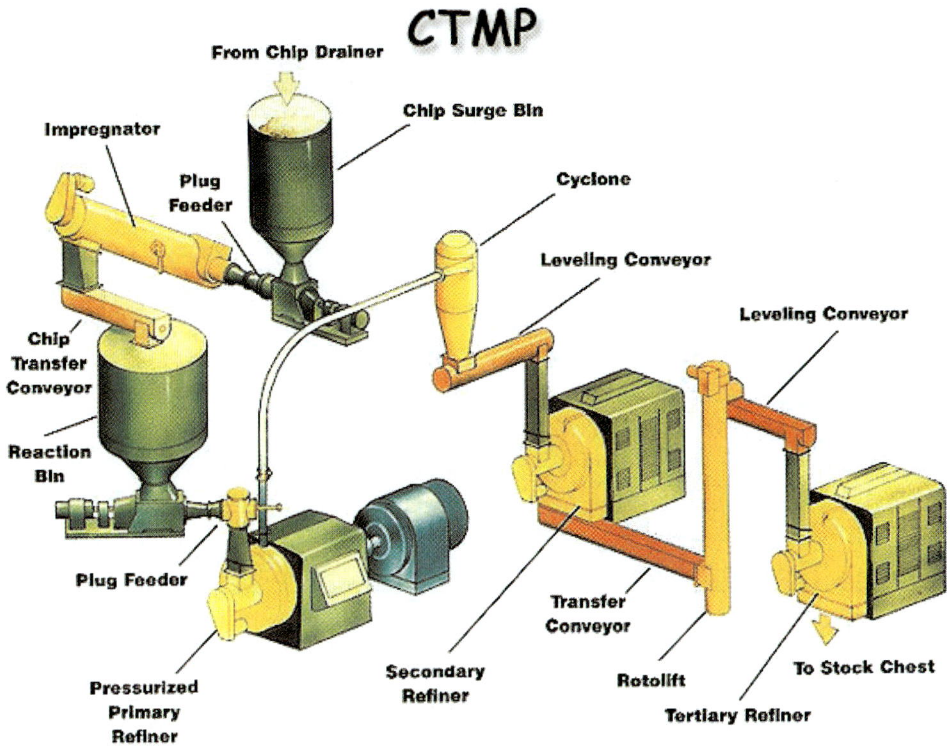

Figure 18: CTMP process flow diagram

In both the TMP and CTMP processes, the refining energy is not uniformly distributed throughout the surface of the fibers. This "latent" energy causes the fibers to twist, kink and curl. These fibers need to be relaxed through heat, agitation and time in a "latency chest" before further processing can be done. The pulp is then screened and thickened and sent to storage chests before it is used on the papermachine. CTMP can also be bleached after thickening, usually with peroxide in a separate operation. The resulting pulp, called BCTMP, is not quite as bright as the highest-brightness kraft pulps, but it is still very high-quality pulp, and it is cheaper than kraft pulp.

Sulfite cooking liquor is not the only cooking chemical that can be used to make Chemi-mechanical pulp. One variation of Chemi-mechanical pulping is to use caustic and peroxide in the initial pretreatment stage. This process is called alkaline peroxide mechanical pulping or APMP. Note that we have left out the T in the designation. That's because chip refining is done at atmospheric pressure. APMP is interesting because it does not require a final-stage bleaching sequence to obtain high brightness. So equipment costs are lower than BCTMP, and energy costs are 25% to 30% lower.

BCTMP and APMP are popular pulping processes. One big reason is that bleached chemithermomechanical pulp is cheaper to produce than kraft pulp. But that's not the only advantage. In comparison to kraft pulping, no chemical recovery plant is required, so equipment costs are much lower. Kraft mills require large capital investments, which can only be justified for large mills producing 1,000 tons per day or more of pulp. On the other hand, BCTMP and APMP plants can be much smaller to meet specific mill needs. Moreover,

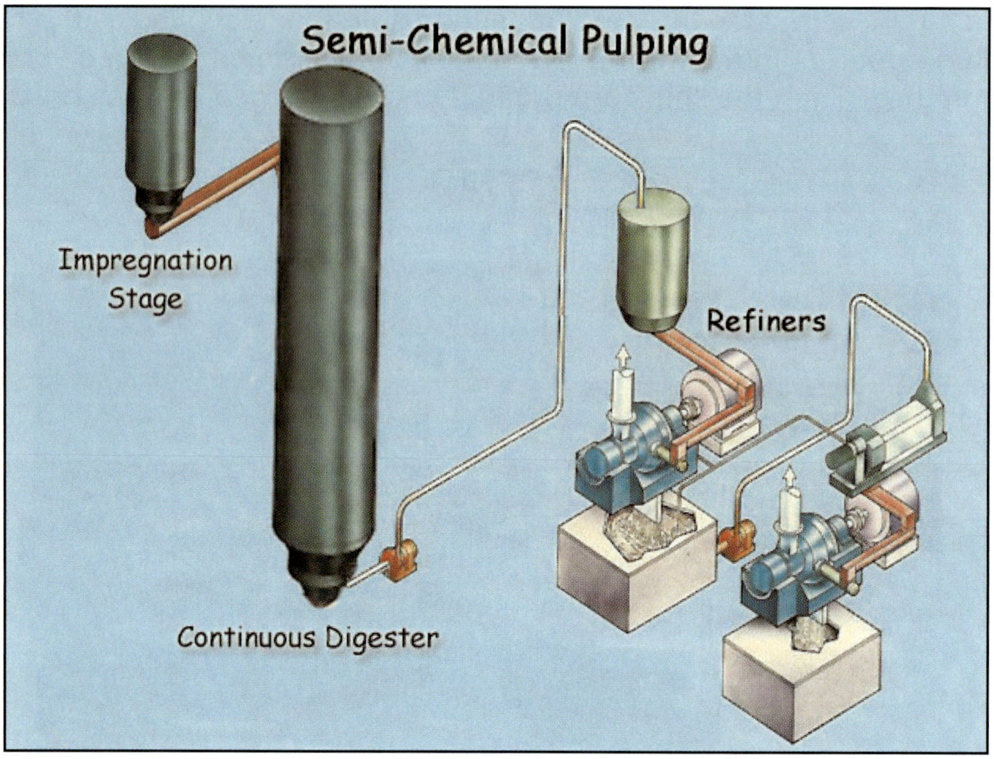

Figure 19: Flow diagram for semi-chemical pulping process

bleached chemithermomechanical pulps have some interesting properties that make them useful in papermaking. One advantage is that the pulp is bulkier than kraft pulp, which means it can be used as an inner layer for multi-layered grades of paper. Bulkier pulp means that the sheet is thick for any given amount of fiber, so it may be possible to reduce the total fiber content of the sheet and maintain the same sheet stiffness. BCTMP also has very good opacity, which improves the printing properties of the final paper. One disadvantage is that the pulp is weaker, so depending on the grade of paper, BCTMP might be mixed with kraft pulp to give the strength necessary in the final paper.

To this point, we have discussed using relatively mild chemical treatments. But what if we used even stronger chemical treatments? Now we have entered the realm of Semi-chemical pulping, which can also be considered high-yield chemical pulping. With stronger chemical treatment of the chips, the yield starts to drop because some of the lignin is being removed. One of the most common Semi-chemical pulping processes is the neutral sulfite semi-chemical or NSSC process (Figure 19).

NSSC was developed over 50 years ago to allow the use of hardwood. Most NSSC pulp is used to make corrugating medium, which is the inner fluted ply in a corrugated box. NSSC pulp is high-yield, around 75%, and is also very strong and stiff. It is also dark brown, so it is not suitable for bleached grades of paper. The NSSC process normally uses sodium sulfite and sodium carbonate as the cooking chemicals. The chips are cooked at 160 to 185 degrees Celsius (320 to 365 degrees Fahrenheit) for 30 minutes or more. Some NSSC mills use other cooking chemicals such as kraft green liquor or sodium carbonate and sodium hydroxide. Once the chips have been cooked, the fibers are separated by refining.

6.6 SUMMARY

Fiber can be isolated from wood by chemical or mechanical processes or a combination of both. Alkaline pulping using the kraft process is the most common chemical method used, but it converts less than 50% of the wood into usable fiber. The oldest and simplest mechanical pulping method is the stone groundwood process. Here, logs are pressed against a stone to yield short, impure fibers suitable for making newsprint. Yields are high (around 95%), but the fibers are weak and unstable, that is, they easily turn yellow in sunlight. A more sophisticated mechanical method that processes wood chips through disc refiners was developed in the 1960s and is known as refiner-mechanical pulping (RMP). An upgrade to this basic mechanical process using steam to soften the chips and force them through the refiners was soon developed and is called thermomechanical pulping (TMP). Both RMP and TMP pulps produce pulps that are measurably stronger than SGW pulp and somewhat more stable. Yields are still in the 90% range.

More recently, other significant upgrades to mechanical pulping were introduced as the Chemi-mechanical (CMP) and chemithermomechanical pulping (CTMP) processes. These hybrid processes use a mild chemical pretreatment that further softens the wood chips and produces a brighter, stronger, more stable pulp. There are many variations of this basic process such as BCTMP and APMP. Yields are still good, averaging around 85%.

An even more rigorous chemical pretreatment of the chips is used to make Semi-chemical pulps (as opposed to the milder treatment for Chemi-mechanical pulps described above). These pulps are almost exclusively used to produce corrugating medium, the fluted material between the inside and outside surfaces of cardboard.

Mechanical, Chemi-mechanical and semimechanical pulps are generally much less expensive to produce than chemical pulps, both because of their high yields and because of the much lower capital costs to construct pulping facilities. Because huge amounts of energy are required to produce them, it is necessary to have an inexpensive source of power available and to use heat recovery as an integral part of the process. The pulps are generally weaker than chemical pulps, but print well because of the high bulk associated with the pulp produced. The processes used to produce them are generally environmentally friendly. Because of the inherent instability of these pulps, they are best utilized in throwaway products or as additives to chemical pulp for non-critical uses where archival requirements are not important. All of the reasons noted here make the various mechanical processes the most popular ways to produce pulp in today's highly competitive economy.

6.7 REVIEW QUESTIONS

You will know that you have understood the materials in this module if you can easily answer the following simple questions:

1. What are some of the differences between mechanical and chemical pulps?
2. How is groundwood produced?
3. What characteristic does pressurized groundwood (PGW) have that makes it a superior pulp compared to stone groundwood (SGW)?

4. What is the basic process used to produce all of the refiner-mechanical and Chemi-mechanical pulps?
5. What is the primary difference between refiner-mechanical and Chemi-mechanical pulp?
6. What advantages do the refiner-mechanical and Chemi-mechanical pulping processes have over groundwood?
7. Why is heat recovery so important in all of the mechanical pulping processes?
8. What are BCTMP and APMP pulps, and why are they popular choices for the manufacture of many paper products?
9. What is the basic Semi-chemical (not Chemi-mechanical) pulping process?
10. What is the NSSC process?

Chapter 7

PULP BLEACHING

OBJECTIVES

At the conclusion of this section, the student will be able to:

- ***Explain how bleaching adds value to pulp and why it is necessary.***
- ***Explain brightness, yield, selectivity, and pulp quality.***
- ***List common bleaching chemicals and bleach sequences.***
- ***List the main equipment components in a bleach stage.***
- ***List important environmental concerns related to bleaching.***
- ***Describe the difference between ECF and TCF bleaching.***

7.1 KEY TERMS

AOX - "Adsorbable Organic Halides". This test is typically performed on effluents from mills that bleach pulp. It is used to indicate how much chlorinated organic material was formed as side products of bleaching. AOX is synonymous with organochlorine.

Bleach Demand - Amount of bleach chemical required for bleaching of a specific pulp. Bleach demand is determined by performing tests such as kappa number and brightness on pulp.

Bleaching Sequence - A series of bleaching stages that contribute to the overall brightening of pulp.

Bleaching Stage - A process carried out to remove lignin and brighten the pulp using one or more chemicals, elevated temperature, time, and sometimes higher pressure.

BOD (Biological oxygen demand) - A measurement typically performed on mill effluent that indicates how much oxygen the effluent would consume if discharged directly into natural receiving waters. Most bleached kraft pulp mills have operating permits that limit how much BOD the mill can release to the surrounding environment.

Brightness - A test performed on pulp that measures the reflection of visible light under standard conditions.

Carbohydrates - A synonym for polysaccharides.

Carryover - The amount of material carried over with the brownstock into the bleach plant. Usually, more carryover means more bleaching will be needed.

Cellulose - A long-chain molecule made up of glucose sugar units joined together in a linear fashion. Cellulose is a carbohydrate and makes up 40-50% of most wood types.

Color - A measurement typically performed on mill effluent that indicates how dark the effluent is expected to be once released into the environment.

COD - A measurement done on a liquid that indicates the "chemical oxygen demand" present. When there is high carryover of COD in the brownstock entering the bleach plant, more bleach chemical will be required.

Consistency - The amount of dry pulp in 100 g of slurry. A pulp of 5% consistency is made up of 5 g dry fiber and 95 g water or filtrate.

Chelant - A chemical that deactivates metals in pulp so that hydrogen peroxide can react well during bleaching.

Chemical Pulp - Pulp produced using a process to separate the wood fibers chemically by dissolving and removing most of the lignin. Because the wood fibers are separated with minimal mechanical damage to the fiber, chemical pulps are typically stronger than mechanical pulps.

Delignification - The process of chemically removing lignin from wood or pulp. This is mainly accomplished in a digester. Lignin removal is also the main purpose of oxygen delignification.

ECF Bleaching - Bleaching that uses chlorine dioxide, but not chlorine gas.

Filtrate - Liquid obtained from a pulp washer that contains most of the spent delignification or bleaching chemicals and dissolved solids. Bleach filtrates are usually sent to the waste treatment plant, while brownstock and oxygen delignification filtrates are sent to the black liquor evaporators.

Hemicellulose - A molecule much smaller than cellulose made from several different sugar units. Hemicellulose is a carbohydrate and makes up 20-30% of most wood.

Lignin - A natural polymer in wood that holds fibers together. Lignin makes up 20-32% of wood, depending on the wood type. Softwoods (e.g. pine, spruce, fir) have more lignin than hardwoods (e.g. oak, maple, poplar, aspen) and so are more difficult to pulp and bleach.

Market Pulp - Pulp of any type that is produced for sale on the open market or on contract to specific customers. It is then used as raw material for paper and other products.

Mechanical Pulp - Pulp produced through mechanical action to separate the wood fibers. Because the wood fibers become damaged through the mechanical action of the refiners, mechanical pulp is typically not as strong as chemical pulp.

Mixer - A piece of equipment used to mix different materials together, such as steam, chemicals, oxygen, and pulp.

Polysaccharides - A term used to describe the sum of cellulose and hemicellulose.

Reactor - A piece of equipment used to carry out a controlled chemical reaction.

Selectivity - Relates to the fact that different bleaching chemicals react differently with carbohydrates than with lignin. "Better selectivity" means that the bleach chemical reacts more with lignin than with carbohydrates.

TCF Bleaching - Bleaching that does not use any chemicals that contain a chlorine (Cl) atom. Such sequences rely entirely on oxygen-based bleaching chemicals such as ozone, hydrogen peroxide, peracetic acid, and molecular oxygen.

TOC - A measurement done on a liquid that indicates the "total organic carbon" present.

Yield - The percentage of pulp recovered after one or more processing steps.

7.2 INTRODUCTION

Many paper products, such as food board, printer and magazine paper must be bright white. To make bright pulp, pulp must be bleached to remove the brown color (Figure 1).

Figure 1: Pulp bleaching

If the kraft pulp being produced is going to be bleached, then pulping is allowed to proceed until 90% or more of the lignin originally in the wood is removed. In other words, the pulp is cooked to a low kappa number. You may recall from the kraft pulping chapter that the kappa number is a measure of the remaining lignin and lower numbers mean less lignin. Unfortunately, the small amount of lignin remaining in the unbleached pulp gives it a characteristic light brown color. Why not simply cook the pulp until all the lignin has been removed? First, the pulp yield would drop dramatically. Secondly, excessive cooking will damage the pulp fibers, resulting in poor strength of the final paper. But the real reason is, it just doesn't work chemically. Bleaching is the way to remove residual lignin selectively, causing very little damage to the fibers while making them bright.

There is more than one way to make pulp whiter. Bleaching and brightening may sound similar but they are quite different chemically. Bleaching actually removes lignin and is used to increase the brightness of chemical pulps. Brightening converts chemical groups in lignin to forms that do not darken pulp, thereby making it appear whiter. Brightening is used for mechanical or chemimechanical pulps that still contain lots of lignin. The disadvantage is that it is not permanent - yellowing occurs after exposure to light.

Pulp is bleached or brightened to improve the whiteness of the final product. Not all paper products require the same level of whiteness. For example, you can go to your local office supply store and buy economy-grade copy paper with a brightness of 85, or you could choose a high-grade copy paper with a brightness of 100. What do these numbers mean? Typically, brightness compares the amount of light reflected from a paper sample with that from a known standard, usually magnesium oxide. The higher the percent reflectance, the brighter or whiter the paper - and the more bleaching the pulp received.

7.3 CHEMICAL PULP DELIGNIFICATION AND BLEACHING

7.3.1 Oxygen Delignification

One stage in pulp manufacture, called oxygen delignification, has become much more commonplace during the 1980s and 1990s. Oxygen delignification, sometimes called "O_2 Delig" for short, is sometimes called bleaching, but it is really part of the "brown" part of a pulp mill. This is because the oxygen delignification exists mainly to remove lignin, rather than to brighten the pulp. Figure 2 is a schematic of the oxygen delignification process.

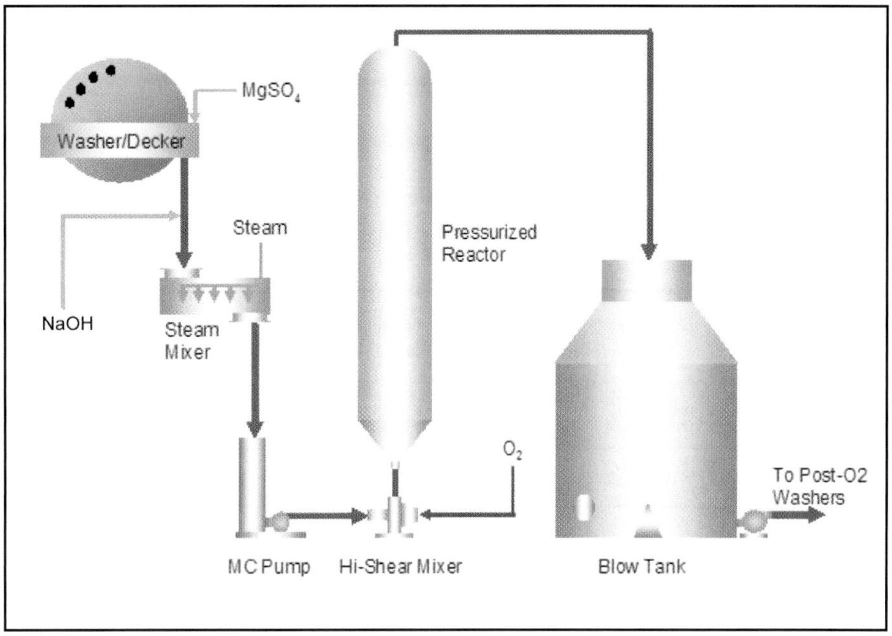

Figure 2: Oxygen delignification process

Oxygen is used in an alkali solution under pressure to delignify unbleached pulp. A major advantage of oxygen delignification is that it has better selectivity than kraft cooking. It can remove up to fifty percent of the lignin without significant damage to the pulp fiber. Consequently, bleaching costs are much lower for mills with oxygen delignification than for mills without it.

The equipment can be costly since a pressurized vessel is usually required and good post-O_2 washing is desirable. Oxygen delignification also dramatically decreases the load on effluent treatment facilities. This is because most of the TOC and COD in bleach filtrate comes from lignin, and less lignin enters a bleach plant after oxygen delignification.

Oxygen delignification is done at medium (10-18%) or high consistency (>20%). Medium-consistency systems are far more common. Recent systems have two consecutive reactors rather than one in order to improve the efficiency and selectivity of O_2 delignification. Pulp washing, both before and after an O_2 delignification reactor, is important, and high-efficiency washers such as twin-roll wash presses and multi-stage vacuum drum filters are often used.

7.3.2 Common Bleaching Stages and Sequences

When chemical pulps are bleached, the bleaching is always done using a sequence of bleach stages. Common bleaching stages along with their bleaching symbols are:

D - acidic stage using chlorine dioxide (ClO_2)

E - alkaline extraction stage using sodium hydroxide (NaOH)

P - alkaline stage using hydrogen peroxide (H_2O_2)

A - acidic hydrolysis stage commonly using sulfuric acid (H2SO4) or D-stage filtrate

Z - acidic stage using ozone (O_3)

X - stage using enzymes to make pulp easier to bleach

Q - stage using a chelant to deactivate metals that interfere with peroxide bleaching

Typical bleach sequences use 3-5 bleach stages. Table 1 lists typical conditions for a bleach sequence to produce a market pulp of 90 ISO brightness.

	O_2 Delignification	D	Eop	D	E	D
ClO_2 on pulp, %		3.0		1.0		0.3
NaOH on pulp, %	2.0	1.5		1.0		
O_2 on pulp, %	1.0	0.5				
H_2O_2 on pulp, %			0.3			
Final pH	11	3.0	11	3.5	11	4.0
Consistency, %	10	10	10	10	10	10
Temperature, ºC	110	50	70	70	70	70
Time, minutes	60	45	90	180	60	180
Kappa number in	30	16				
Kappa number out	16		5			0
Delignification, %	47		69			
Brightness, % ISO	35	45	60	82	80	90

Table 1: Typical bleach sequence conditions to product 90 ISO brightness pulp

The sequence in the table above would be written and spoken: D(EOP)DED. This is a so-called ECF or "elemental chlorine free" sequence, meaning no chlorine gas (Cl_2) is used as a reagent.

A common configuration for a three-stage bleach sequence is shown in Figure 3. In this schematic, a brief description is included of how each stage serves its unique purpose.

Pulp bleaching has changed considerably over the last 200 years, but the part that has remained is the use of some form of chlorine (Cl). Because of its atomic nature, Cl can be found in many chemical forms that behave in completely different ways, such as molecular chlorine (Cl_2), sodium hypochlorite (NaOCl), and chlorine dioxide (ClO_2). For example, sodium chloride (NaCl), also known as table salt, is an important part of our body's electrolytic balance that makes every muscle motion possible. Too much of it, however, and table salt becomes lethal. Such is the case also with bleaching chemicals. When chlorine gas (Cl_2) was the predominant bleaching agent, many more chlorinated organic by-products were formed during bleaching. These proved hard to break down during wastewater treatment and in the natural environment. With the advent of ClO_2 as the bleaching agent of choice, much less environmentally objectionable matter is formed during bleaching, since ClO_2 reacts in very different ways with pulp than does Cl_2. Today, ClO_2 is the primary bleaching agent of choice worldwide due to its cost effectiveness, selectivity, ability to produce a high-quality bleached pulp, and low environmental impact. Table 2 shows the evolution of bleaching processes.

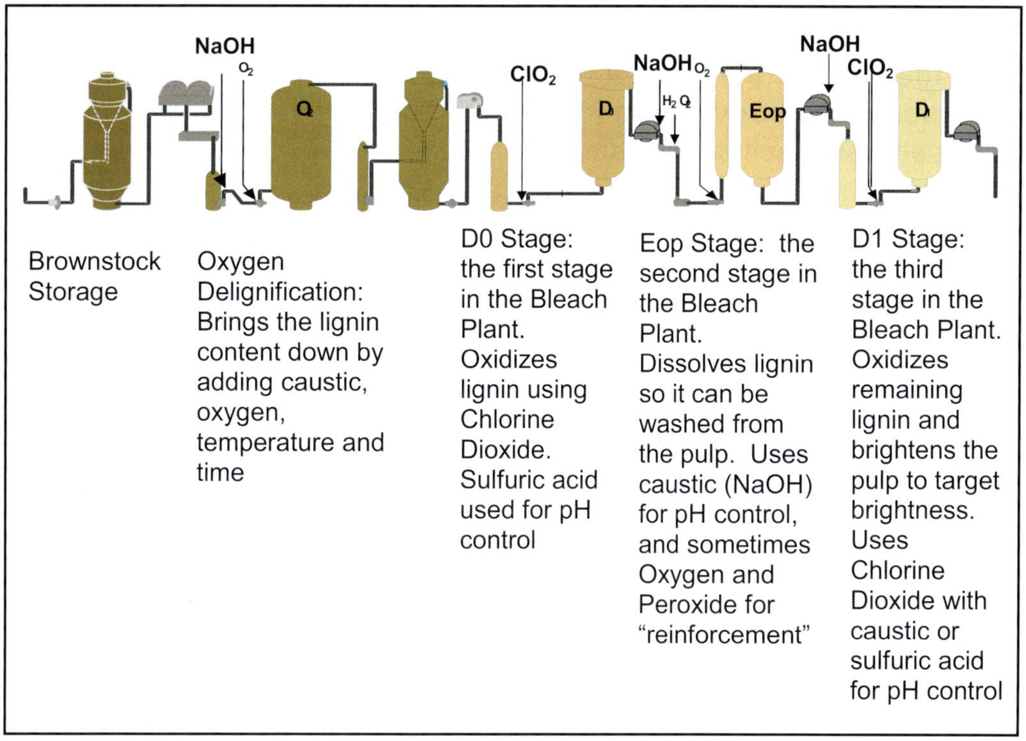

Figure 3: Typical three-stage bleach sequence

Year	TAPPI Designation	Informal Designation
<1800	Air + sun	Air + sun
1800	H, HH	H, HH
1900	HEH	HEH
1930	CEH	CEH
1950s	CEHD	CEHD
"	CEHDED	CEHDED
1970	(C+D)EHDED	CDEHDED
1980s and '90s	(C+D)(EO)DED	CDEoDED
"	(D+C)(EO)DED	DCEoDED
"	(D+C)(EOP)D(EP)D	DCEopDEpD
"	D(EOP)D(EP)D	DEopDEpD

Table 2: Evolution of bleaching process

7.3.3 Bleaching Equipment

Each bleaching stage is a particular combination of chemistry, equipment, and process conditions. In a conventional bleaching sequence, the equipment is the most similar aspect of the stages. Each stage usually includes 5 main components: 1) a **steam mixer** to heat the pulp suspension with direct steam, 2) a **pump** to transport the pulp, 3) a **chemical mixer** to combine pulp with the bleaching chemicals, 4) a **tower** to allow time for the bleaching to occur, and 5) a **washer** to separate the spent bleaching solution from the bleached pulp and minimize carryover to the next bleaching stage. Figure 4 illustrates the equipment in a bleaching stage.

Like everything in a pulp mill, it's not quite that simple in reality - there are lots of types of stock pumps, chemical mixers, bleaching towers, and washers.

A modern bleach plant producing high-brightness pulp - let's say a D(EOP)DED sequence - commonly has a low-consistency ClO_2 first stage with one or two chemical mixers, an upflow tower, and a rotary vacuum drum filter for pulp washing. The ClO_2 comes to the mixer at a concentration of about 10 grams per liter in cold water. The pulp suspension, at 3.5% consistency, has already been heated to about 50ºC. The mixing is essentially instantaneous. The pulp and ClO_2 go to the retention tower to react for about 45 minutes, after which the pulp is washed. Subsequent alkaline extraction and chlorine dioxide stage conditions vary widely depending on several factors such as overall bleach sequence configuration, pulp species, and target brightness.

7.3.4 Bleaching Chemicals

The amount of bleaching chemical needed depends on which chemical, which stage in the sequence, and what kind of pulp. With ClO_2, for instance, progressively less is used as you go along the sequence. In a D(EOP)DED case on softwood pulp, the amount of ClO_2 used in each D stage may be roughly 3%, 1%, and 0.3% applied on pulp (i.e. 3% = 60 lb ClO_2 per ovendry ton of pulp). On the same pulp, the Eop stage might use 1.5% NaOH on pulp, supplemented by 0.5% O_2 and 0.3% H_2O_2. The E stage will probably use 1% NaOH on pulp. The exact numbers will depend on the bleaching job to be done, always trying to use each bleaching reagent as efficiently as possible.

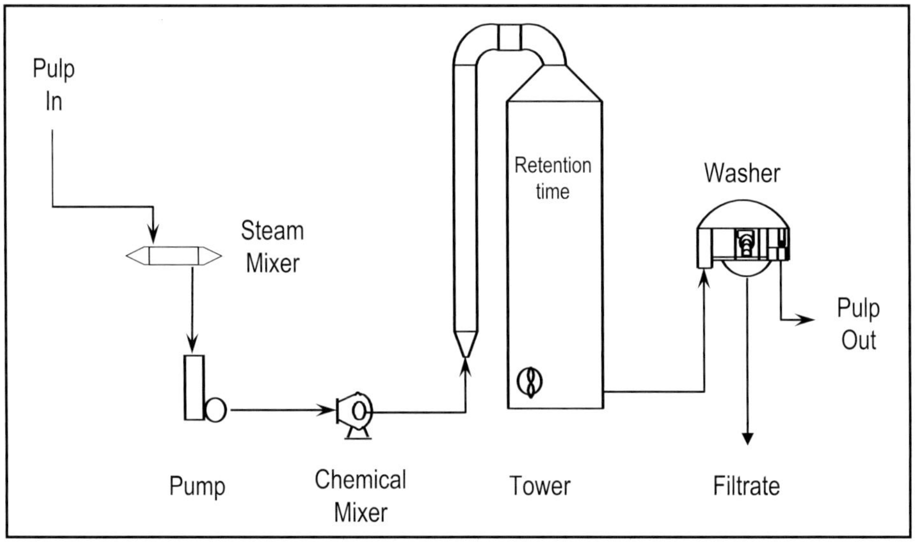

Figure 4: Typical equipment in a bleaching stage

The objective of bleaching is to remove the residual lignin

from the unbleached pulp, making it white. Although many bleaching reagents have been used over the years, ClO_2 is by far the preferred one worldwide because it is very selective in dissolving residual lignin without degrading the polysaccharides - the cellulose and hemicelluloses which make up the most important part of the pulp fiber. Along a bleaching sequence, ClO_2 stages are always interspersed with caustic extraction stages. This alternating pattern of acidic and alkaline stages is able to break down the ever-smaller amounts of residual lignin, ultimately dissolving all of it so that it can be washed out of the pulp.

7.3.5 Increasing Brightness During Bleaching

Pulp brightness doesn't increase steadily across a bleaching sequence. Rather, it goes up only modestly in O_2 delignification, substantially in the first and second ClO_2 stages, and modestly again in the final D-stage. The caustic extraction stages don't chemically whiten the pulp - they are there to dissolve and remove the lignin which has already reacted with the ClO_2 in the previous stage. Figure 5 shows how brightness commonly develops during bleaching.

Market pulps are normally taken to a final brightness of 90% ISO or higher, at which point they appear completely white. But bleached pulp being made into photocopy paper in an integrated pulp and paper mill may be bleached to 85 percent brightness instead.

7.3.6 Bleach Plant Filtrate Flow

Just as we saw earlier in brownstock washing, the bleach filtrates flow counter-currently in a bleach plant, which means opposite to the flow of pulp. But in this case, there are really two filtrate flows. In one of them, filtrate from the final D-stage washer is used as shower water on the second D-stage washer, and its filtrate is used on the first D-stage washer. Some of the acidic filtrate from the first D-stage washer is used to dilute and control the consistency of the pulp coming to the bleach plant from pulp storage. The rest exits the bleach plant and goes to effluent treatment.

The filtrate flows of the two caustic extraction stages are connected in a similar fashion, with the final alkaline filtrate leaving the first E-stage washer and going to effluent treatment. Unlike the yellowish acidic filtrate stream, this one is distinctly brown - it's that lignin again. In addition to dissolved lignin, both streams contain significant amounts of chloride ion, which negatively affects the chemical recovery system in a kraft mill. An effluent treatment plant has a dual purpose - to react biologically with the dissolved organics to reduce BOD, and to remove the suspended solids. Then the treated effluent can be released safely to the receiving water, typically a river or an ocean, without upsetting the natural oxygen balance of the receiving waters.

7.3.7 Bleached Pulp Quality

Now let's change our perspective a little, and look at the fiberline of a bleached kraft mill in a product quality sense.

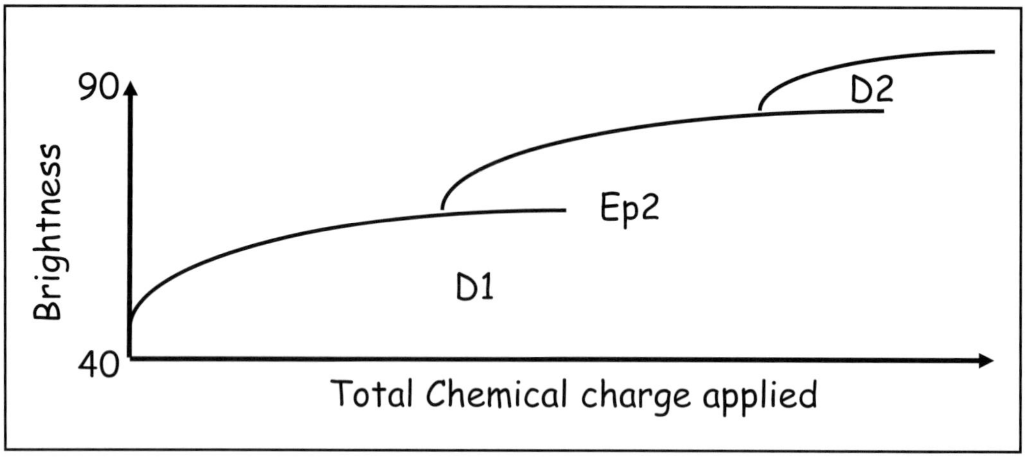

Figure 5: Development of brightness during bleaching

7.3.8 Pulp Strength

Kraft pulping is known for its ability to produce the strongest woodpulp - no other pulping process comes close. So just how effective is a typical fiberline in this respect? In overall physical strength, softwood kraft pulp actually declines along a fiberline, failing in pulping itself to meet the potential strength inherent in the chip supply, and then losing more strength along the rest of the process line.

Fortunately, the losses *after* kraft cooking are usually modest. And despite this strength loss, there's still no better way to make strong pulp for papermaking. In hardwood kraft mills, the pulp strength picture is better, but not perfect. Chip quality, appropriate pulping equipment, and good operating practices are key to producing strong kraft pulps.

7.3.9 Pulp Cleanliness

Other quality aspects of bleaching include cleanliness - the absence of dark particles caused by bark, pitch, or inorganic contaminants - and the minimization of extractives -organic compounds originating in the wood and still present in the pulp. Any fiber bundles have to be converted into bleached pulp fibers. Sometimes metal ions must be removed via chelation. For some uses, pulp quality specifications are particularly rigorous. For example, bleached pulp used to make photo-grade paper cannot have any background radioactivity.

7.3.10 Pulp Yield

Just as yield is important when pulping the wood, the same is true in bleaching the pulp. Bleached yield refers to the amount of dry bleached pulp obtained compared to the amount of dry wood used to make the pulp. Because less than 10 percent of wood's original lignin content remains in the unbleached pulp, and also because O_2 delignification and ECF bleaching are quite selective for lignin removal, the yield loss in bleaching is low.

7.3.11 Evolving Bleach Sequences

During the late 1980s and early 1990s, scientists discovered that under certain bleaching conditions using Cl_2, a class of highly toxic chemicals could be formed in tiny amounts as a result of unwanted side reactions. These compounds were collectively referred to as "dioxins" and "furans", and demonstrated high toxicity to laboratory animals. The discovery of these compounds triggered an unprecedented amount of research and development to discover new ways to bleach pulp that would not produce these toxic compounds. Fortunately, scientists determined that simply by eliminating the use of Cl_2 and ensuring that dioxin- and furan-free defoamers were used, the formation of chlorinated dioxins and furans could be just about eliminated. This is how the term "Elemental Chlorine-Free", or ECF, bleaching, came into use.

On a parallel path, however, some researchers made significant progress in identifying ways to bleach pulp using other bleaching chemicals that did not contain any atomic chlorine. Such sequences were collectively termed "Total Chlorine-Free", or TCF, and relied on chemicals such as oxygen (O_2), hydrogen peroxide (H_2O_2), ozone (O_3), and peracetic acid (PAA, or CH3COOOH). Some bleached pulp manufacturers have been able to create competitive advantage through TCF bleaching. Others, however, have found that bleaching costs and losses in pulp strength can often be significantly higher than for ECF bleaching, while customers were not willing to pay extra for such products.

Some supporters of TCF bleaching suggest that TCF bleaching is inherently more environmentally friendly than ECF bleaching. However, there does not appear to be consensus among scientists, regulators, and manufacturers either way. Clearly, ECF bleaching is the method of choice for bleached pulp manufacturers worldwide.

7.4 MECHANICAL PULP BLEACHING

After chemical pulping dissolves most of the lignin in the original wood, the bleaching processes we have examined can remove most or all of the remaining lignin to produce white pulp. But high-yield mechanical pulps are another matter - they cannot be bleached in this way because they contain too much lignin. Even so, such pulps can be brightened to improve their whiteness and permanence, making them suitable for many grades of paper. Remember, though, that bleached or brightened mechanical pulps almost never reach the same high brightness levels as chemical pulps, and that the brightening effect is reversible when the final paper product is exposed to sunlight.

The most common brightening agents are sodium hydrosulfite and hydrogen peroxide (H_2O_2). A two-stage peroxide sequence can be used to raise the brightness of chemithermomechanical pulp to 85 percent or higher. Both hydrosulfite and peroxide attack the chemical groups that cause darkening and render them colorless. Unlike chemical pulp bleaching, lignin is retained in brightening.

7.5 BLEACHING AND THE ENVIRONMENT

Just as bleaching processes have changed dramatically during the twentieth century, so too has the impact of bleaching on the environment. Historically, the biggest impact a pulp mill has had on bodies of water downstream is that of depleting oxygen from the water. This is because small amounts of lignin and carbohydrates are discharged from bleaching and from liquor losses in the mill. When lignin and carbohydrates are present in rivers and lakes, naturally occurring micro-organisms break down these materials into smaller compounds they can use for food. To do so, these micro-organisms need oxygen. As they consume this oxygen, there is less oxygen left over for fish and vegetation. In extreme cases, bodies of water downstream from a pulp mill many decades ago had little if any fish or aquatic life. The laboratory test that tells how much oxygen the effluent will require in the environment is called BOD (for biological oxygen demand).

A second way that pulp mills have affected bodies of water downstream is due to the color of their effluent. Color in effluent blocks sunlight from penetrating deep into the river or lake water. This blockage can dramatically change the ecosystem if the effluent has a lot of color, and if there is not enough river or lake water to mix with the effluent.

In the 1970s and 1980s, many mills worldwide installed biological waste treatment systems to reduce the BOD of mill effluent. These systems commonly remove more than 90% of effluent BOD, but are not generally designed to remove effluent color.

7.5.1 Chlorinated Organic Compounds and AOX

With the discovery of chlorinated dioxins and furans in bleach filtrates during the 1980s and 1990s, public outcry led to regulations worldwide to reduce the amount of chlorinated organic material (measured by AOX) in bleached pulp mill effluent. This has contributed to the growth and continued improvement of oxygen delignification and ECF bleaching worldwide. While scientists have generally not found AOX to be dangerous to the environment, the industry has continued to reduce AOX discharge during the 1990s and 2000s, mainly by designing and operating bleached pulp mills more efficiently.

7.5.2 U.S. Environmental Protection Agency (EPA) Cluster Rule

The U.S. EPA Cluster Rule became the official federal regulation for the pulp and paper industry when it was published in the *Federal Register* on April 15, 1998. The Cluster Rule got its name because it sets limits on several important environmental parameters to ensure air and water quality. The limits were developed with considerable input from industry and the public. They are based on what is widely known as the "Best Available Technology" that is economically achievable for industry. In a kraft bleach plant, this meant ECF bleaching, elimination of hypochlorite, use of hydrogen peroxide-enhanced extraction (Eop), and high-shear mixing. While the U.S. EPA has no jurisdiction outside of the U.S., many other countries and the European Union have environmental guidelines similar to the Cluster Rules.

To reduce emissions of dangerous chemicals to the air, the Cluster Rules also require that vent gases from bleaching be treated to remove any hazardous air pollutants (HAPs) that might be present.

States can set their own regulations to ensure air and water quality. These regulations must be at least as stringent as the federal regulations. For example, the state of Maine requires that any fish that live downstream of a bleached pulp mill cannot have chlorinated dioxin or furan levels that are higher than those in fish that live upstream from the mill.

7.6 SUMMARY

Pulp bleaching is an important process that adds value to the pulp. It is carried out by several bleaching stages that usually alternate from acidic to alkaline conditions depending on the bleaching chemical used. Common bleaching chemicals include oxygen, chlorine dioxide, hydrogen peroxide, and ozone. Because lignin is colored, the main purpose of bleaching is to remove lignin and make the pulp bright.

Bleaching of kraft and chemical pulps is necessary because excessive cooking in the digester will excessively weaken the pulp.

In the 1980's and 1990's, a class of toxic chemicals called chlorinated dioxins and furans were found to be formed as a by-product of bleaching. However, mainly through elimination of chlorine gas as a bleaching agent, these compounds are no longer formed. In 1998, the U.S. EPA introduced the "Cluster Rule" to minimize the environmental impact of bleaching, mainly by requiring that U.S. mills use the "best available technology" that is economically achievable.

Mechanical pulps can also be bleached, but usually not to the extent that is possible with chemical pulps.

7.7 REVIEW QUESTIONS

1. List two ways that bleaching adds value to pulp.
2. Explain why it doesn't make sense to cook the pulp so that all of the lignin is removed in the digester.
3. Describe the main purpose of oxygen delignification.
4. Explain why bleaching costs are different for mills with oxygen delignification compared to mills without oxygen delignification.
5. Give the full name for the following bleaching stages and state whether they are acidic or alkaline bleaching stages: D, E, P, Z.
6. True or False: The main purpose of the final D stage in a bleach sequence is to remove lignin.
7. True or False: Chlorine gas is a more environmentally friendly bleaching chemical than chlorine dioxide.
8. How much chlorine gas is used in an ECF bleaching sequence? 0%, >50%, 100%
9. True or False: It is common to use hydrogen peroxide to "reinforce" a chlorine dioxide bleaching stage.
10. True or False: A steam mixer is commonly used to mix bleaching chemicals with the pulp.

11. What are the 5 main components of a typical bleaching stage?
12. True or False: Under normal conditions, yield loss is low during bleaching.
13. True or False: Fiber strength increases as a result of pulping and bleaching.
14. Explain what is meant by ECF bleaching.
15. How is ECF bleaching different from TCF bleaching?
16. True or False: Most bleached kraft pulp in the world is made using ECF bleaching.
17. Explain how mechanical pulp bleaching is different from kraft or chemical pulp bleaching.
18. What is AOX and why is it important?
19. What is BOD and why is it important?
20. What is meant by the "Cluster Rule" and when did it go into effect?

7.8 BIBLIOGRAPHY

1. Handbook of Pulp and Paper Terminology: A Guide to Industrial and Technological Usage, Gary A. Smook, 1990, Angus Wilde Publications.
2. Pulp Bleaching - Principles and Practice, Carlton W. Dence and Douglas W. Reeve, 1996, TAPPI Press.
3. 2006 TAPPI Improving Bleaching Operations Course Notes, 2006, TAPPI Press.
4. 2001 TAPPI Pulp Bleaching Course Notes, 2001, TAPPI Press.

Chapter 8

PAPER RECYCLING

OBJECTIVES

At the conclusion of this section, the student will be able to:

- ***Demonstrate knowledge of the terminology used for paper recycling.***
- ***Identify several wastepaper grades and types and describe some of the complexities involved with procuring and transporting wastepaper.***
- ***Demonstrate knowledge of the types of contaminants as well as contaminant sources, properties, and possible means of removal.***
- ***Identify general types of repulpers.***
- ***Identify several types of cleaners and screens and identify contaminants that can be removed by these devices.***
- ***Identify different deinking methods.***
- ***Describe clarifiers and sludge treatment units and discuss how water is cleaned and reused in recycling operations.***

8.1 KEY TERMS

Bale - A large mass of paper product pressed together and usually bound with wires.

Brightness -The reflectivity of pulp or paper for specific blue light under standard test conditions. Generally measures the whiteness of the pulp or paper.

Centrifugal Cleaner - A conical pressure vessel that separates heavy and light contaminants from fibers by spinning the slurry. Separation is by centrifugal force.

Clarification - Removal of suspended solids from water, usually by flotation or settling.

Consistency -The amount of pulp in an aqueous (water) slurry, expressed as a percentage of the total mass.

Corrugated Box - A shipping container made of two layers of linerboard with an inner layer of fluted corrugating medium.

Contaminant - Any material that must be removed from paper during recycling, including dirt, ink, stickies, and debris.

Deinking - Processes used to remove ink and other filler materials from printed recycled paper.

Detrashing - Using specialized equipment to remove large contaminants from recycled paper.

Dirt - Undesirable contaminating material in pulp and paper which is visible.

Dispersion - A process to loosen ink particles and other contaminants from recycled paper fibers and to break them down in size so they are not readily visible.

Flotation - A process where solids are floated with air bubbles. Usually contaminants are floated to be skimmed and removed.

Ink - Pigments and dyes applied to paper products. These must be removed during paper recycling.

Junk Tower - A large-diameter vertical cylinder, located beside the pulper, which is used to collect large contaminants from wastepaper. Normally emptied by means of a grapple.

Kraft Pulp - Pulp made by an alkaline pulping process that uses sodium hydroxide and sodium sulfide to dissolve lignin and release the cellulose fibers.

Linerboard - A heavyweight paper that is used as inner and outer facing material in making corrugated containers.

Makeup - Water added to a system to replace the water lost through usage or purging.

MOW - Short for "mixed office waste", a grade or type of post-consumer recycled paper.

OCC - Short for "old corrugated containers," a grade or type of post-consumer recycled paper.

ONP - Short for "old newsprint", a grade or type of post-consumer recycled paper.

Off-spec - Paper that does not meet the specification ordered by a customer.

Opacity -The ability of paper to resist the transmission of light through it.

Post-Consumer Waste - Paper which has served its intended end use and then been discarded.

Pre-Consumer Waste - Waste or scrap paper created in a manufacturing or converting process. Generally a higher grade of wastepaper than post-consumer waste.

Pressure Screen - Equipment used to remove contaminants from pulp by allowing fibers to pass through holes or slots in a screen basket while rejecting material that is larger or less flexible than the fibers.

Printability - A general term that refers to the properties of paper that result in good print quality.

Pulper - A machine designed to break up fiber and disperse dry pulp and recycled paper in water. May consist of a tub with an agitator or else a rotating drum.

Ragger - A long rope of baling wire, plastics, and other contaminants which is allowed to grow in a pulper and which is slowly and continuously drawn out of the pulper to remove the contaminants.

Runnability - Refers to how well the pulp furnish to a machine forms a sheet on the wire and passes through the drying and finishing sections.

Screening - Process used in paper recycling to separate fibers and contaminants by size by passing material through screen holes or slots.

Secondary fiber - Fiber stock that comes from a paper recycling operation.

Sludge -The solids removed from the effluent of a pulping, ercycling, or paper process

Stickies - Suspended particulate contaminants consisting of hot-melt glues, films, tapes, and other adhesives or polymers that, if left in the pulp, cause spots in the final product and accumulated debris on the papermaking equipment.

Surfactant - A chemical (such as soap) that stabilizes a suspension of oils, greases, and inks. Allows ink to be trapped in foam.

Urban Forest - A term used to refer to all the paper fiber which is available for recycling in urban areas.

Virgin Pulp - Pulp produced by chemical or mechanical means which has not yet been made into paper or paperboard.

Washing - Process used in recycling to remove contaminants by water displacement.

8.2 INTRODUCTION

Today, in North America, we recover almost 50% of all the paper we use. Some of this recycled fiber is shipped overseas, but recovered fiber still makes up close to 40% of the fiber used to make paper and paperboard products in North America.

Paper recycling has good support from the government, consumers, and paper companies, but it is economics that drives recycling. In countries or regions that have a ready supply of virgin fiber from trees and wood, recycling often will not happen because it does not make good business sense. From the papermaker's standpoint, wastepaper can be a good fiber source which is competitive in price to kraft pulp made at the mill and cheaper than purchased virgin pulp. Many mills use recycled fiber to increase their fiber supply to the paper mill instead of building new kraft pulp mill capacity. Smaller recycled paper mills can be located in industrial parks near the "urban forest" and near the mill's customers.

From the standpoint of the consumer, recycling keeps good paper from being thrown into our landfills. Recycling also takes the pressure off our forests to supply all the fiber we would otherwise need. We have more trees now than we did 70 years ago, and recycling is an integral part of sustainable forestry.

8.3 TYPES OF WASTEPAPER

Recycled paper is classified as either post-consumer or pre-consumer waste. Post-consumer waste is paper that has been converted into a product, served the purpose for which it was intended, and then been discarded by the user. A good example of post-consumer wastepaper is old newspapers, or ONP. Pre-consumer waste is paper created in a manufacturing or converting operation. An example of pre-consumer wastepaper is box plant trimmings, sometimes called double-lined kraft or DLK, which is waste generated in the manufacture of corrugated boxes. For the most part, post-consumer waste has more contaminants and is considered a lower grade of recycled fiber.

There are a lot of different grades or classifications for recycled papers, but let's look at some of the more common ones.

- Mixed Paper: paper of varied quality such as office waste (MOW), boxboard cuttings and mill wrappers.
- Old Newspapers (ONP)
- Old Corrugated Containers (OCC)
- Pulp Substitutes: mainly consists of unprinted paper and board, such as tabulating cards, trimmings, shavings and cuttings.

Pulp substitutes are utilized directly and are generally fully recovered. Other wastepapers must be cleaned to remove contaminants.

8.4 WASTEPAPER PROCUREMENT

To enable wastepaper to be utilized efficiently as paper furnish, it is necessary to sort the collected material by quality grade. This is either done at the source or by the collection agency or wastepaper dealer. Mills monitor their incoming wastepaper to ensure that minimum quality standards are adhered to.

Wastepaper recovery methods have changed in recent times and most ONP and OCC now comes from curbside collection or other recycling programs operated by cities and counties. Thus municipalities play a major role in wastepaper collection and supply (especially in the case of ONP). Wastepaper dealers play a minor role in collection and sorting. In the case of OCC, many chain stores have a compactor that bales OCC, and many municipalities collect OCC from homeowners. In fact, around 75% of all corrugated boxes are collected, baled, and sold as OCC.

Figure 1: Photographs of watepaper bales

The easiest paper product to recycle is used corrugated boxes. Over 95% of the goods we use daily are shipped in corrugated boxes, and 30% of the paper we produce in North America is used to make corrugated boxes. No other grade of paper is so widely recycled.

8.5 CONTAMINANTS IN WASTEPAPER

Depending on the supplier and the source of collected wastepapers, there may be a wide variation in the type and level of contamination. Contaminants can be defined as anything that you do not want in the final product. All extraneous constituents such as dirt, rocks, sand, and tramp metal are considered contaminants. Other common problem contaminants are hot-melt glues, waxes, tape adhesives, and other "stickies" as they are called in the paper mill.

Step one in the recycling process is therefore to inspect and sort the incoming bales of recycled paper for quality. The cleaner the incoming wastepaper, the easier it is to recycle it into clean new paper. Really bad bales can be rejected, and other bales can be separated based on the suspected level of contaminants. Then the cleanest bales can be mixed with more contaminated ones to keep the level of contaminants under control.

Simple separation and/or dispersion of contaminants such as plastics, laminates, adhesives, and glue is sufficient for pulps that are used for the inner plies of multi-ply paperboard or for corrugating medium. But for manufacture of printing grades, more extensive removal of contaminants, including deinking, is necessary to make the pulp suitable for papermaking. Newsprint is a good example. Newsprint makes up about 15% off all paper and paperboard produced in North America, and 50% of all old newspapers, referred to as ONP, are recycled, making this the second most commonly recycled paper grade behind OCC.

8.6 REPULPING

Once bales are sorted, they are fed into the pulper (or repulper) so the wastepaper can be broken up into individual fibers and made into a pulp slurry.

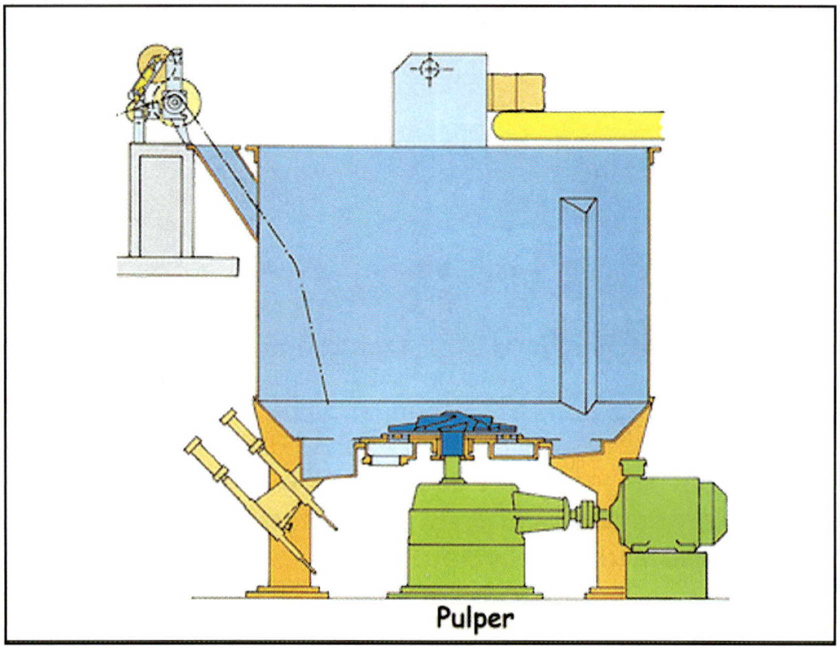

Figure 2: Batch pulper

A conventional pulper looks a lot like a very large kitchen blender. There is a tub with a rotor in the bottom. Bales are mixed with hot water and the action of the rotor breaks apart the fibers in the paper. Depending on the contaminants to be removed, additional chemicals such as dispersants and surfactants are added to the pulper. Temperature, consistency, retention time, and chemistry are controlled to achieve good dispersion of the fibers and detachment of contaminants.

Pulpers may be set up to run in batch or continuous mode. A batch pulper is shown in Figure 2. Most pulpers are operated in continuous mode wherein the bales of waste paper are continuously fed into the pulper along with hot water, and the pulp is continuously extracted through the bottom of the pulper through a screen plate that is usually located underneath the rotor.

Unfortunately, there are other things besides fiber that get into the pulper and need to be removed. A ragger (Figure 3) is an older, but still effective, method of contaminant removal. The rotation of the material in the pulper forms a spun rope from bailing wire, plastic pieces, and other material. The ragger slowly pulls this rope of rejects out of the pulper and new material continues to collect on the tail that is still in the pulper.

Other large contaminants are removed on a continuous basis through a junk tower and detrasher, as shown in Figure 4. During the pulping process, it is important to try to keep from breaking the contaminants into small pieces. For example, it is far easier to remove a whole tennis shoe than it is to take out little pieces of rubber, vinyl, and cloth.

Drum pulpers (Figure 5) work at high consistency on a continuous basis. Bales enter one end of the drum and the rotating action lifts the paper and repeatedly drops it. Fibers are separated, but the contaminants are left in large pieces, which make them easier to screen out. As the pulp moves through the drum, it enters the screening zone, where dilution water helps flush the fibers through the screen plate openings. Rejects are discharged through the end of the drum intact and fiber-free, so losses are very small.

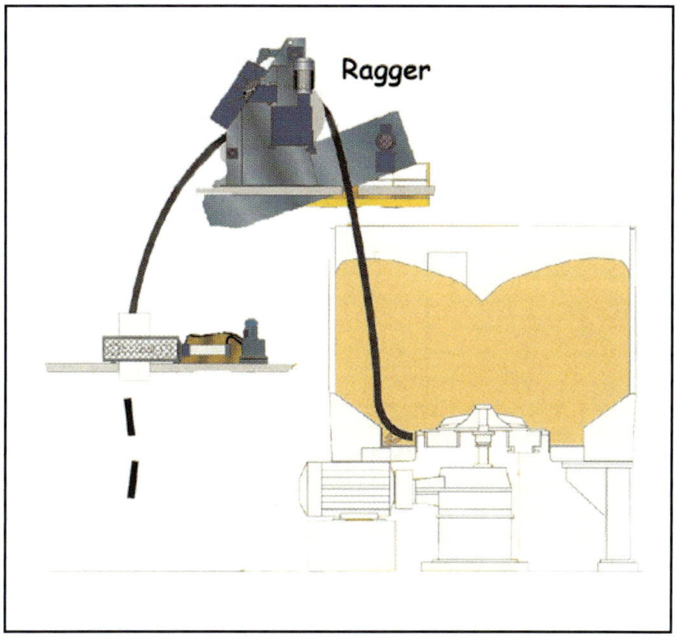

Figure 3: Ragger

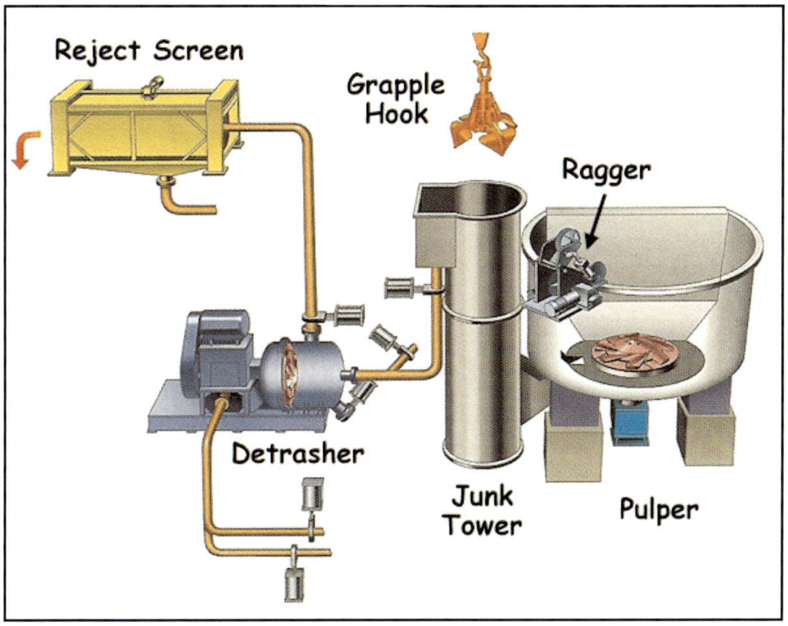

Figure 4: Schematic of pulper, junker, detrasher

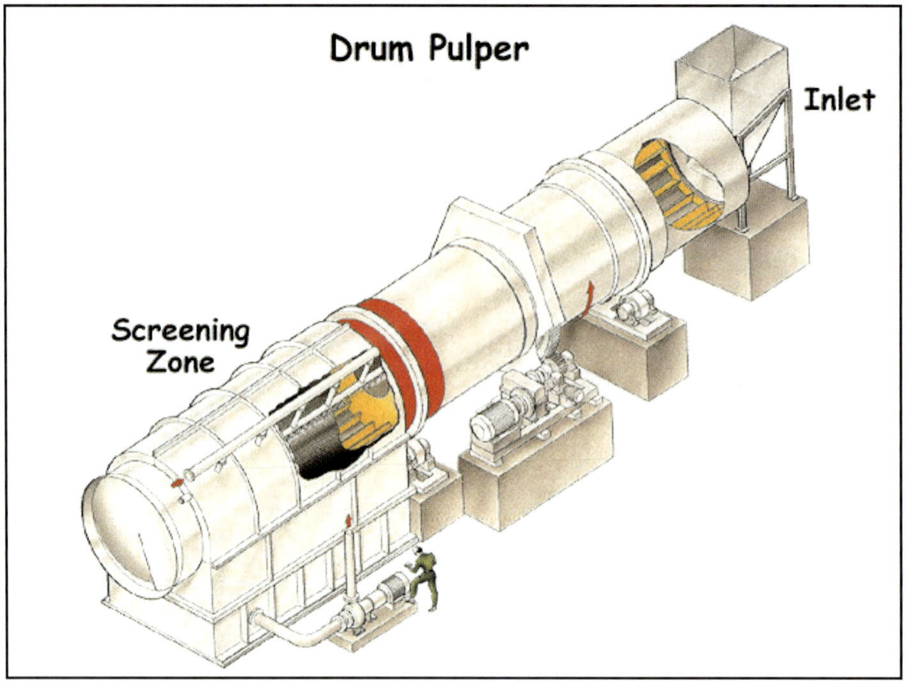

Figure 5: Drum pulper

8.7 CLEANING AND SCREENING

Once the pulp has left the pulper, the pulp slurry is usually pumped through high-consistency centrifugal cleaners to remove larger heavy materials such as staples, rocks, and glass. The stock enters the top of the centrifugal cleaner at an angle tangential to the body of the cleaner. Inside the cleaner, the stock spins around like a cyclone or tornado, and the centrifugal forces push the heavier contaminants to the sides of the cleaner. Heavy contaminants then work their way to the bottom of the cleaner where they are discharged, and the good fiber exits through the top of the cleaner. Figure 6 illustrates how a high-consistency centrifugal cleaner works.

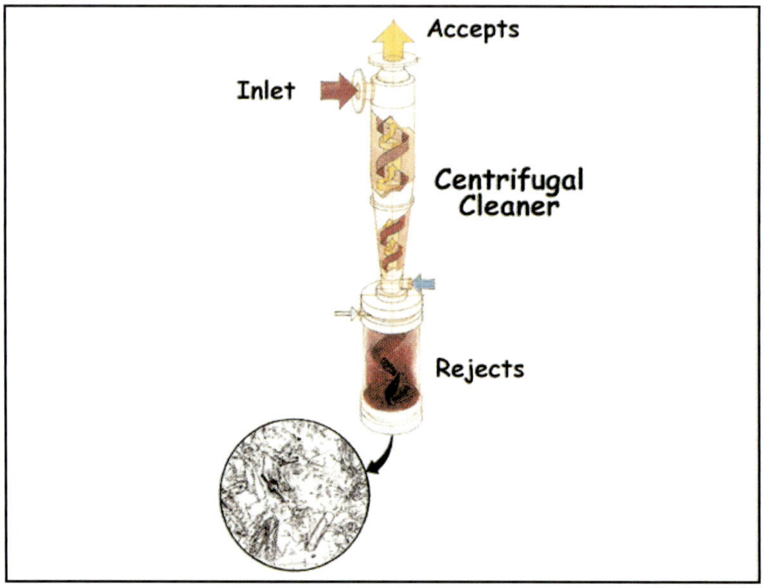

Figure 6: High density centrifugal cleaner

Depending on the type of pulper and the final end use for the pulp, the stock may be coarse-screened to remove larger contaminants such as any remaining fiber flakes and larger pieces of plastic and wood, and then fine-screened to remove smaller contaminants such as stickies and bits of dirt. Modern pressure screens (Figure 7) consist of a screen basket with holes or slots and a rotor with some type of hydrofoil to keep the

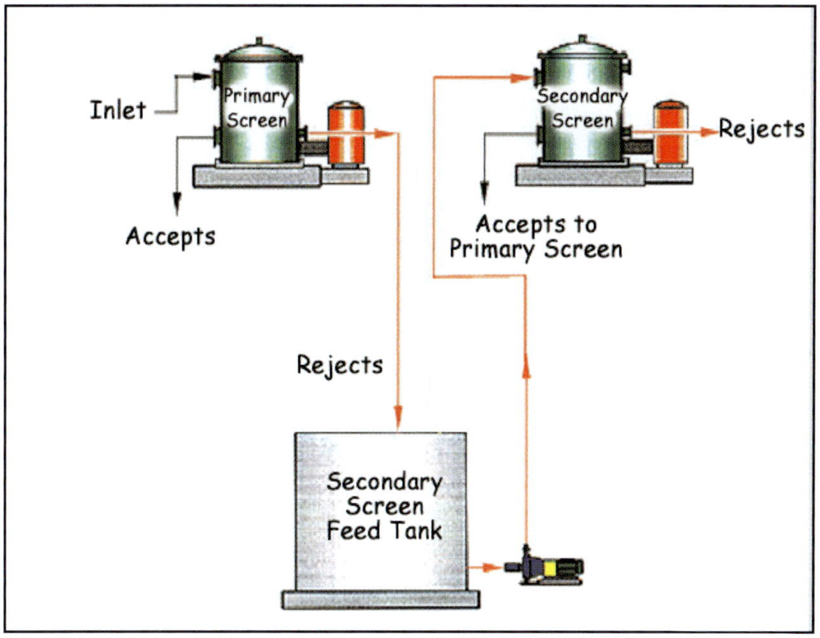

Figure 7: Pressure screen flow diagram

screen holes or slots from plugging. In the example shown, the hydrofoils produce a pressure pulse that pushes and then pulls fiber from the openings.

Both coarse and fine screening is usually done in multiple stages arranged so that a minimum of fiber is lost from each system. In coarse screening, a number of different styles of screens can be used. Most screens are centrifugal, as shown in Figure 8, meaning that the stock flows from the inside of the cylinder to the outside. Another type of coarse screen uses the opposite, or centripetal, flow pattern where the stock flows from the outside to the inside of the screen cylinder.

Fine-slotted screens are perhaps the most important step in making high-quality stock where sheet appearance is of primary importance. Screening out contaminants with fine slots is a method for controlling stickies, so a lot of attention needs to be paid to this step in the process. Doing so will result in much better pulp quality and lower overall operating costs. Usually arranged in multiple stages, the fine screening operation uses low-pulse rotors that don't push stickies through the screen plate openings. Modern screens use slots that range in size from 0.006" to 0.008" so that extremely fine debris is removed efficiently with little loss of good fiber. Figure 9 displays two types of screen baskets.

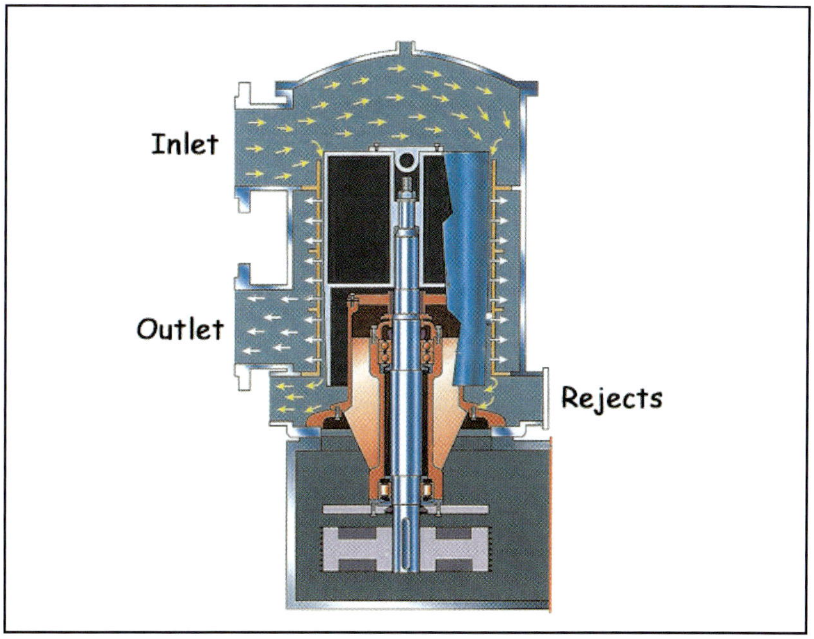

Figure 8: Centrifugal screen

High-density centrifugal cleaners (discussed earlier) remove the larger heavy contaminants from pulp at 4-5% consistency. However, the pulp also needs to be cleaned at low consistency to remove smaller contaminants. Low-density cleaners (Figure 10) work at a stock consistency around 1%. The action inside the cleaners is the same as in a high-consistency cleaner. The centrifugal action separates the heavy particles such as sand and moves them to the outside walls of the cleaner where they move to the bottom of the cleaner and are discharged. Good fiber is discharged through the top of the cleaner. Some cleaners are set up to work in reverse (Figure 11), so that light contaminants such as expanded packing foam collect in the center core of the cyclone and are discharged through the top of the cleaner. In this way, low-consistency centrifugal cleaners can be used to remove both lightweight and heavyweight contaminants.

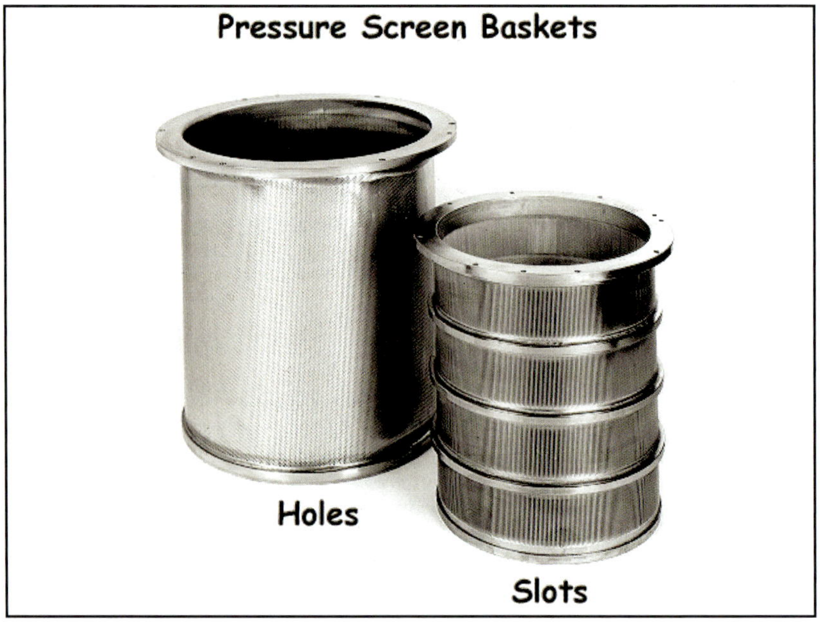

Figure 9: Screen baskets

8.8 DEINKING - FLOTATION AND WASHING

For brown grades such as linerboard, corrugating medium, and grocery bags, screening and

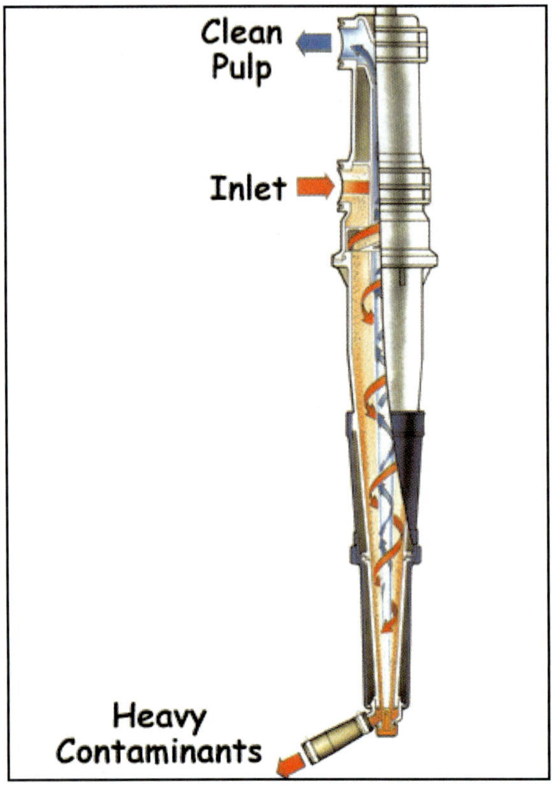

Figure 10: Low-density cleaner

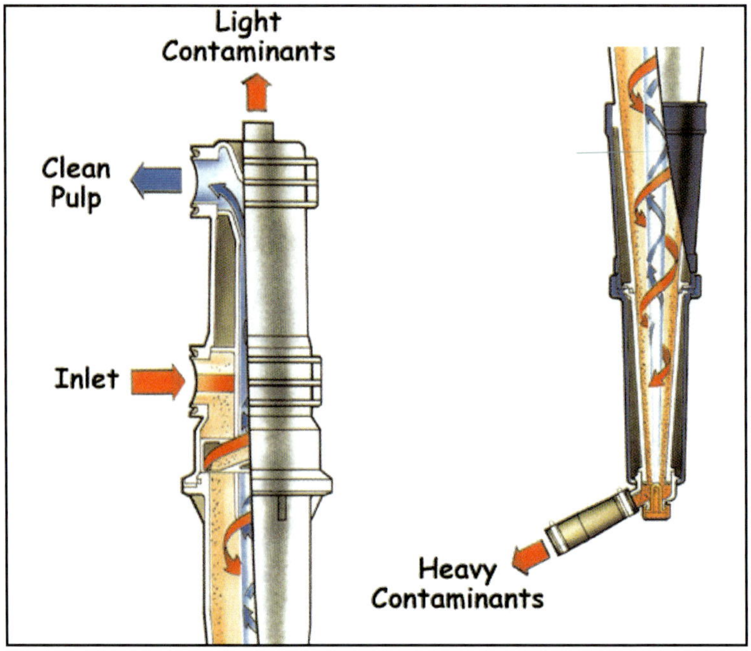

Figure 11: Reverse cleaner

cleaning the stock is sufficient. However, if you want to make recycled pulp for a grade which has brightness and printability requirements, you need to do all the pulping, cleaning, and screening that was described earlier, but you also need two more steps: deinking and brightening.

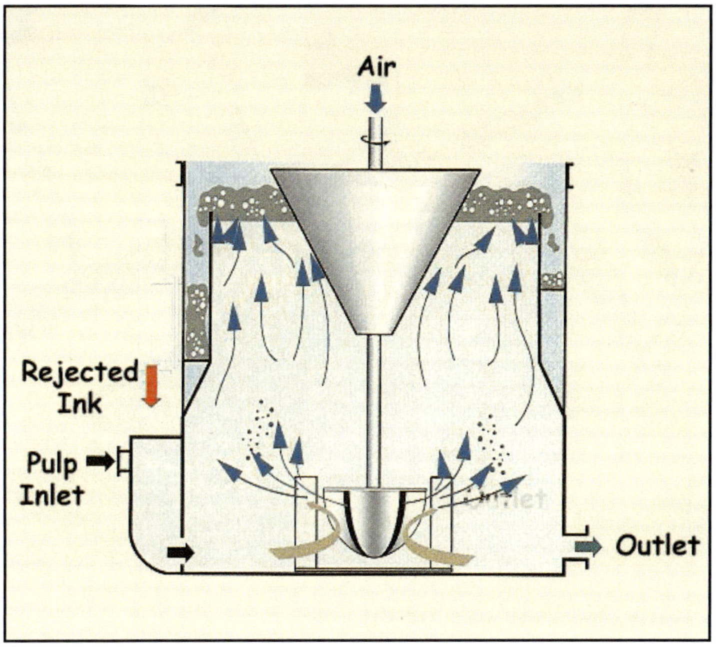

Figure 12: Rectangular flotation cell

Printing ink makes up 1% to 2% of the total weight of ONP. That ink, if allowed to stay with the pulp, would result in low brightness as well as numerous ink specks in the final paper. So, what is the best way to remove the ink from the fiber? The options are to try to wash it out of the fibers or to try to float it off the fibers, and some processes use both techniques.

Deinking really begins in the pulper. Pulpers that are designed for deinking can run in batch or continuous mode (depending on whether they are tub or drum pulpers) and at high consistency, usually between 18% and 20%. Pulp at these consistencies looks like bread dough. The fiber-to-fiber rubbing action, along with added chemicals (such as detergents, dispersants, foaming agents and caustic soda), helps loosen the ink from the fiber. Once pulp has been processed in the pulper, it has to be diluted from high consistency to around 4 or 5% so the pulp slurry can be pumped to the next steps in the process.

After cleaning and screening to remove as much ink and other contaminants as possible, the pulp is further diluted to about 1% consistency and sent to the deink flotation cell. Here, air in the form of very small bubbles is mixed with the pulp slurry. The ink particles attach themselves to the rising air bubbles. The foam that forms on the surface of the stock contains the ink particles. This foam is skimmed off to separate the ink from the good fibers. Flotation deinking is usually followed with more cleaning and screening to remove additional contaminants. In flotation, the critical operating parameters are controlling the bubble size (usually by the design of the cell's air injectors), getting the ink particles to the right size so that the bubbles can carry them away, and controlling water and foam chemistry.

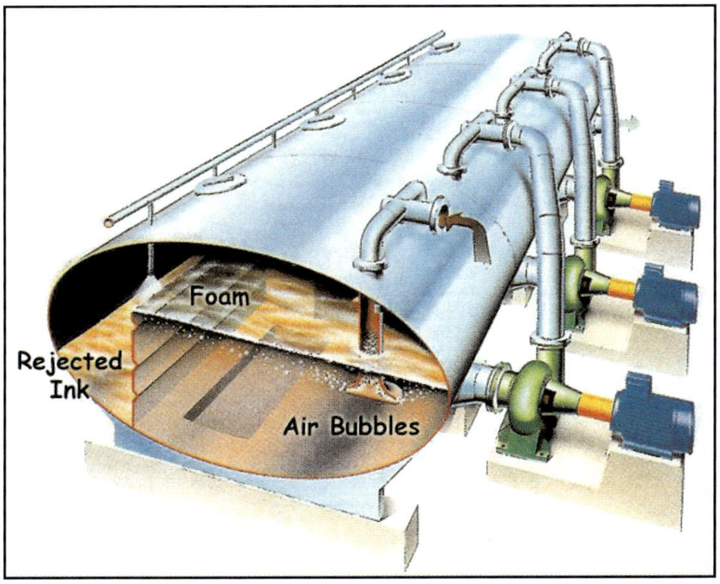

Figure 13: Multi-injector elliptical cell

Figure 14: Wash deinking

That is why in some deinking plants there are two stages of flotation: one before dispersion and one after. The one before dispersion removes the softer inks that are loosened in the pulper, and the one after dispersion removes the harder inks that need to be loosened from the fibers by the high-consistency rubbing action of the dispersion step. Figure 12 shows a rectangular flotation deinking cell while Figure 13 shows a multi-injector elliptical cell.

Dispersion machines can be either kneaders or refiners. In kneaders, the pulp is fed between two large screws rotating in opposite directions that impart a lot of shear to the pulp. In refiner-type dispersers, the pulp is fed between two discs, one of which rotates and imparts shear to the pulp. In either case, the effect is to apply a lot of power to the high-consistency pulp slurry so that the fibers rub together, forcing the ink to break down into sub-visible particles that can be removed by the next steps in the process.

Ink can also be removed from pulp by washing (Figure 14). Chemicals are mixed with the pulp, which cause the ink to emulsify in much the same way that you would wash oil or grease from your hands. Once the ink is soluble, it can be washed from the fiber on sidehill screens or other washing equipment. It usually takes several stages to wash the ink from the pulp effectively. For pulps that will be made into tissue, washing is extremely important

Figure 15: Dissolved air clarifier

because it also removes papermaking additives that do not contribute to fiber bonding strength, like clay fillers used in printing and writing papers. Tissue made from secondary fiber cannot have any of this material in the sheet, so washing is a must in the deink plant making tissue as a product.

Washing efficiency depends mostly on how much ink-laden water is removed from the pulp and the subsequent clarification steps. Inky water from the washing step must be clarified in a dissolved air clarifier, or DAF (Figure 15). A DAF acts much like an ink flotation cell, using dissolved air to remove all the solids from the water so it can be recycled back into the system.

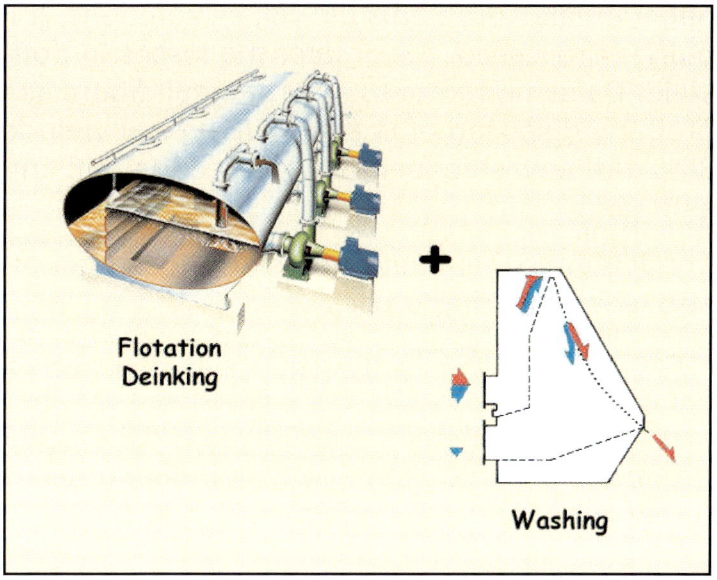

Figure 16: Flotation/washing combination

Modern hybrid deinking systems, as shown in the schematic in Figure 16, combine flotation and washing to achieve an overall improvement in pulp brightness.

8.9 BLEACHING AND BRIGHTENING

Some recycled pulps, especially mixed office waste or MOW, which might be used for high-brightness copy paper or other publication grades or even tissue, may require even more processing.The pulp might go through a bleaching stage using oxygen and hydrogen peroxide, similar to the final bleaching stages used for kraft pulps (Figure 17).

The recycled pulp or secondary fiber is then thickened to 10% or higher consistency and stored in the high-density chest until needed on the paper machine.

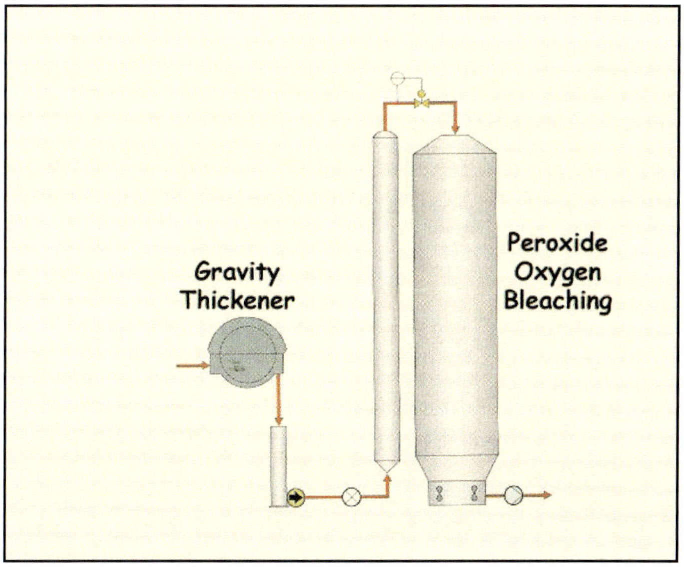

Figure 17: Bleach schematic

8.10 WASTE DISPOSAL

Recycled paper is a prime source of fiber for the paper industry. We have reached our goal of recovering 50% of all paper used in the United States! However, recycling of paper creates problems as well as solutions. It is important to remember that the more the pulp is processed, the greater the losses in both contaminants and fiber, and the lower the yield. Deinking removes a lot of small fibers as well as ink particles, and this sludge must be properly disposed of in a landfill. Typical yield for deinked pulp is around 70%. That means 30% of the wastepaper that entered the mill ends up being hauled back out of the mill as sludge.

In addition to yield loss, fibers also lose strength each time they pass through the recycling and manufacturing process.

Processing chemicals and energy are expensive, and recycled pulp may end up being as expensive as, or even more expensive than, higher-quality kraft pulp. From a purely processing standpoint, recycling is most efficient when the wastepaper is reused in paper products which do not have demanding appearance requirements.

8.11 CONCLUSION

By keeping contaminants out of wastepaper right at the start, it is much easier for the mill to recycle the paper in an efficient manner. Once the mill receives the bales of wastepaper, the receiving department must inspect the bales for contaminants and separate the paper according to quality. Most mills use several different types of wastepaper in their furnish, such as OCC, ONP, and MOW. Different types and quality of wastepaper are fed to the pulper in a specific "recipe" so that the quality of the final recycled pulp will be as consistent as possible.

Once the bales of wastepaper have entered the pulper, the fibers are separated by the action of the rotor and the rubbing action of the fibers against each other. The stock is then cleaned and screened to remove as many contaminants as possible. If the pulp is going to be used in a grade of paper such as newsprint, publication papers, or tissue, then the stock must be further processed in a deinking operation followed by a brightening or bleaching operation.

The success of paper recycling depends on the availability of good-quality wastepaper. How carefully we sort the paper we put out for curbside collection ultimately affects the quality of the wastepaper stream and the ability of a mill to reuse the paper efficiently.

8.12 REVIEW QUESTIONS

1. Define secondary fiber.
2. Secondary fiber plants are usually located in areas of __________ ______________.
   Why?
3. Name the basic grades of wastepaper.
4. What do the following acronyms stand for?
   a. ONP
   b. OCC

5. Which type of wastepaper is used the most?
6. Define pre-consumer wastepaper.
7. Define post-consumer wastepaper.
8. List the primary steps in handling wastepaper at a mill.
9. List the six major process steps that may be used in contaminant removal.
10. What are stickies?
11. List common contaminants in wastepaper repulping systems.
12. What is the purpose of deinking?
13. Ink particles are separated from ONP by either ______ or _______ steps or a combination of these.
14. Name typical chemicals used in deinking.
15. How does a flotation cell work for deinking?
16. At what typical consistency does a flotation cell work?
17. What problems are presented by secondary fiber utilization in a paper mill?

ANSWERS TO REVIEW QUESTIONS

Chapter 1 - Wood Fibers

1. What is photosynthesis?

   (A) A process by which green plants turn carbon dioxide and water into sugar.

2. What are some of the differences between hardwood and softwood fibers?

   (D) Hardwood fibers are much shorter and are better for paper printability.

3. What is lignin?

   (B) The glue that holds fibers together in the tree.

Chapter 2 - Pulp Mill Overview

1. What are the two basic processes used to isolate fiber from wood?

   Chemical pulping, and mechanical pulping

2. What are the two kinds of mechanical pulping processes?

   Stone groundwood pulping, and refiner mechanical pulping

3. What are the two chemical pulping processes?

   Kraft and sulfite pulping

4. What are the advantages and disadvantages of mechanical pulping?

   Advantages are higher yield, and bulky with good opacity; disadvantages are weaker fibers and shorter, impure fibers.

5. What are the advantages and disadvantages of chemical pulping?

   Advantages are longer, purer fibers that are stronger; disadvantages are lower yield and harder to print on.

6. What is semichemical pulping and why is it used?

   Semichemical pulping is a hybrid process usually involving a short chemical pulping process that enables the fibers to be separated with less mechanical action than a regular mechanical pulping process. Semichemical pulping is used mainly to produce a pulp stronger than regular mechanical pulp.

7. What are some typical products that use mechanically produced pulps?

   Newsprint and phone books

8. What are some typical products that use chemically produced pulps?

   Books, annual reports, and copy paper

9. What is the biggest use of semichemical pulp?

    Corrugating medium

Chapter 3 - Wood Preparation

1. Explain why the woodyard separates hardwood and softwood chips.

    Softwood and hardwood chips require different pulping and bleaching conditions due to their differences.

2. Why are mills using more purchased chips?

    Because wood has more value as a wood product or building product than as paper. So, pulp mill receives residual chips from wood products facilities, as well as chips from wood chip mills.

3. Explain how a debarking drum works.

    Logs enter a slightly tilted, rotating drum from the top and exit at the bottom. The force of impact of the logs against each other and against the inside shell of the debarking drum causes shearing of the bark off the log as they pass through the drum.

4. List the impacts that longer storage time has on wood.

    Chip moisture loss, insect damage, and less wood extractives such as pitch, tall oil and turpentine.

5. What happens to the bark that is removed?

    Bark is commonly used as a fuel in on-site boilers.

Chapter 4 - Kraft Pulping

1. What does the kappa number measure?

    It measures the amount of lignin in pulp.

2. What are the differences between white liquor and black liquor?

    White liquor is used to cook the chips, and black liquor is what is produced as a result of pulping.

3. What are the two main types of digesters?

    Batch and continuous

4. What is the H-factor and why is it useful?

    H-factor is a number that is calculated from the time and temperature of a kraft cook. It measures the extent of cooking. If the H-factor is too high, then the pulp will be overcooked, and if it is too low, then the pulp will be undercooked.

5. What is the purpose of the disc refiner which treats the pulp when it leaves the digester in a high kappa number kraft pulp mill?

   A disc refiner is usually needed with a higher kappa pulp (e.g., linerboard) to provide mechanical action to separate the fibers.

6. Why is brownstock washing necessary?

   Brownstock washing separates the cooked pulp from the spent cooking liquor. This allows the washed pulp to be further processed, and the spent cooking liquors to be recovered in the recovery process.

7. Describe two types of pulp washers.

   Pressure washers and vacuum washers.

8. Why does brown stock need to be screened?

   Screening lets the good usable fibers pass through the screens while it rejects fiber bundles and other impurities that are not desirable for papermaking.

9. Briefly describe a pressure screen and how it works.

   It is a type of screening that uses an enclosed cylinder where pressure pushes good fibers (accepts) through holes or slots. The holes or slots are kept free of rejects usually by continuous automatic cleaning.

10. Why does pulp need to be thickened before storage in a high density storage tower?

    Because the pulp going into a high density storage tower normally comes directly from the screening process, which is done at low consistency. If low consistency pulp is stored in a high density storage tower, there would be too much water in the tower which would dramatically reduce the storage residence time.

Chapter 5 - Chemical Recovery

1. Calculate the % sulfidity and % activity of green liquor with the following ABC test results:

   A = 5.0; B = 7.5; C = 20.5

   % sulfidity = 2(7.5-5.0)/7.5 = 66.7%

   % Activity = 7.5/20.5 = 36.6%

2. Name different types of brown stock washers.

   Rotary vacuum washers, Horizontal belt washer, Diffusion washers, Pressurized rotary drum washers

3. State the objective of brown stock washing.

The function of the brownstock washers is to separate the black liquor from the pulp fibers.

4. List the main steps of the recovery process in the kraft recovery furnace.

   Evaporates the remaining water in the black liquor, burns the lignin and other wood substances, produces high pressure steam, recovers the cooking chemicals as smelt

5. List the contents of a typical kraft mill black liquor.

   There are two ingredients in black liquor: (1) Lignin and other wood component dissolved during pulping; and @2) cooking chemicals

6. What is the objective of black liquor evaporation?

   In order for the black liquor to be burnt in the furnace, the solids content of the liquor must be raised to above 65%, and this is achieved by evaporation.

7. Describe the operation of a multiple-effect evaporator system with respect to liquor and steam flow.

   The group of evaporators is numbered based on the direction of steam flow. The direction of steam flow is opposite to that of black liquor. The weak black liquor is generally fed both into the Number 6 and Number 5 effects (in a six effect evaporator system), then from both these to the Number 4 effect, and so on. The steam is supplied to the Number 1 effect, and vapor from this effect is used in Number 2 effect, and so on. This countercurrent flow allows the black liquor entering the multiple effect evaporators at around 15% solids to be concentrated to 50% solids or better.

8. Draw the liquor cycle and name the primary operations and liquors.

   See page 53, figure 2.

9. What happens in the causticizing process?

   In the causticizing process, the quick lime is converted to calcium carbonate, called lime mud, and the green liquor is turned into white liquor. This reaction begins when lime is mixed with green liquor in the slaker. The first step is the slaking reaction where quick lime reacts with water in green liquor to form milk of lime (calcium hydroxide). Unreacted lime particles called "grits" are raked from the slaker, washed, and disposed off for landfill. The chemical reaction continues in a series of reactors called the causticizers (usually three or four), where the mixture of green liquor and lime is agitated for several hours to form sodium hydroxide and calcium carbonate.

10. Draw and identify a box diagram of a typical causticizing system, using arrows to show the flow paths.

(This can be simplified as boxes of dissolving tank, green liquor clarifier, slaker, causticizers, white liquor clarifier, mudwasher, mud filter and lime kiln)

11. How is green liquor formed?

The molten smelt from furnace flows into the dissolving tank where it is mixed with weak wash to form green liquor

12. Write the slaking and causticizing reactions.

Slaking reaction:

$$CaO + H_2O \longrightarrow Ca(OH)_2$$

CaO	H2O	Ca(OH)2
lime	water from green liquor	milk of lime

Causticizing reaction:

$$Na_2CO_3 + Ca(OH)_2 \longrightarrow 2NaOH + CaCO_3$$

Na2CO3	Ca(OH)2	2NaOH	CaCO3
sodium carbonate	milk of lime	caustic	lime mud

13. What is the purpose of green liquor clarification?

The purpose of the green liquor clarifier is to remove the dregs. In the green liquor clarifier the dregs settle to the bottom, and the green liquor is removed from the top.

14. What is the purpose of white liquor clarification?

After the causticizers, the mixture contains white liquor and lime mud, and the white liquor and lime mud are separated in the white liquor clarifier, where the mud settles to the bottom and the overflow white liquor is sent to digester.

15. Describe the operation of a lime kiln.

Washed and thickened lime mud is dried and burnt to form quick lime in the lime kiln.

16. What are dregs?

The green liquor is contaminated with bits of carbon and other undissolved material. These suspended solids are called dregs.

17. Match the following liquor properties with its appropriate definition.

a. A　　1. active alkali

b. B　　2. total alkali

c. C	3. effective alkali
d. Na_2S	4. caustic
e. Na_2CO_3	5. sulfide
f. NaOH	6. sodium carbonate
g. grits	7. girls raised in the south
h. dregs	8. unreacted lime
i. white liquor	9. cooking liquor
j. green liquor	10. smelt dissolved in weak wash
	11. carbonaceous impurities in green liquor
	12. sodium oxide

a. A	(3)
b. B	(1)
c. C	(2)
d. Na_2S	(5)
e. Na_2CO_3	(6)
f. NaOH	(4)
g. grits	(8)
h. dregs	(11)
i. white liquor	(9)
j. green liquor	(10)

Chapter 6 - Mechanical and Chemi-Mechanical Pulping

1. What are some of the differences between mechanical and chemical pulps?

   Mechanical pulp contains more lignin than chemical pulp, and so mechanical pulps produce weaker paper. Mechanical pulps have more bulk, stiffness, and opacity than chemical pulps, which can be desirable properties for many grades of paper.

2. How is groundwood produced?

   Groundwood is produced by pressing short logs against a rotating abrasive wheel in which the logs are aligned parallel to the wheel.

3. What characteristic does pressurized groundwood (PGW) have that makes it a superior pulp compared to stone groundwood (SGW)?

PGW pulp is stronger than SGW pulp, because the fibers can be separated with less energy since pressurized steam is applied during PGW pulping.

4. What is the basic process used to produce all of the refiner-mechanical and chemimechanical pulps?

   Disc refining is used to produce all RMP and CMP pulps.

5. What is the primary difference between refiner-mechanical and chemimechanical pulp?

   Chemi-mechanical pulping uses mild chemical action in addition to mechanical action to separate the fibers, whereas refiner-mechanical only uses mechanical action.

6. What advantages do the refiner-mechanical and chemimechanical pulping processes have over groundwood?

   RMP pulping produces a stronger pulp with less fines, as compared to SGW. CMP produces a higher-quality pulp with better brightness relative to SGW.

7. Why is heat recovery so important in all of the mechanical pulping processes?

   Mechanical pulping requires a very high input of energy to operate the equipment that mechanically separates the fibers. Much of this energy gets converted to heat during mechanical pulping. In order for mechanical pulping to be economical, this leftover heat from pulping is used to produce steam for papermaking, helping the paper mill reduce the amount of purchased power they might otherwise need.

8. What are BCTMP and APMP pulps, and why are they popular choices for manufacture of many paper products?

   BCTMP is bleached, chemi-thermo-mechanical pulp, which is a CMP (also called CTMP) that is bleached to increase brightness. BCTMP pulp is one of the most valuable mechanical pulps, and because of its quality and lower cost, it is sometimes used as a substitute for more expensive bleached kraft pulp.

   APMP stands for: alkaline peroxide mechanical pulping. APMP is desirable because pulping is carried out at atmospheric pressure, and the APMP produces a higher brightness pulp.

9. What is the basic semichemical (not chemimechanical) pulping process?

   Semichemical pulping is similar to CMP, but normally relies on a stronger chemical treatment of the pulp than CMP does.

10. What is the NSSC process?

    NSSC stands for Neutral-sulfite semichemical pulping. NSSC pulping is a specific semichemical pulping process that uses sodium sulfite and sodium carbonate as chemicals.

Chapter 7 - Bleaching

1. List two ways that bleaching adds value to pulp.

   Makes it brighter, cleaner and more uniform

2. Explain why it doesn't make sense to cook the pulp so that all of the lignin is removed in the digester.

   Excessive cooking damages the fibers making them weaker, whereas bleaching is more selective than cooking

3. Describe the main purpose of oxygen delignification.

   To remove lignin.

4. Explain why bleaching costs are different for mills with oxygen delignification compared to mills without oxygen delignification.

   Oxygen delignification removes lignin, and so there is less bleaching required in mills with oxygen delignification.

5. Give the full name for the following bleaching stages and state whether they are acidic or alkaline bleaching stages: D, E, P, Z.

   D = chlorine dioxide, acidic; E = alkaline extraction; P = hydrogen peroxide, alkaline; Z = ozone, acidic

6. True or False: The main purpose of the final D stage in a bleach sequence is to remove lignin.

   False, it is to brighten the pulp, mainly by decolorizing any remaining lignin

7. True or False: Chlorine gas is a more environmentally friendly bleaching chemical than chlorine dioxide.

   False, chlorine dioxide makes less AOX and does not make chlorinated dioxins/furans

8. How much chlorine gas is used in an ECF bleaching sequence? 0%, >50%, 100%

   0%

9. True or False: It is common to use hydrogen peroxide to "reinforce" a chlorine dioxide bleaching stage.

   False

10. True or False: A steam mixer is commonly used to mix bleaching chemicals with the pulp.

    False, the chemical mixer accomplishes this function

11. What are the 5 main components of a typical bleaching stage?

    Steam mixer, pump, chemical mixer, bleaching tower, washer

12. True or False: Under normal conditions, yield loss is low during bleaching.

    True, yield loss is 1-3%

13. True or False: Fiber strength increases as a result of pulping and bleaching.

    False; pulping reduces fiber strength significantly, and while bleaching does not dramatically reduce fiber strength, there is usually at least a small decrease.

14. Explain what is meant by ECF bleaching.

    Elemental chlorine-free

15. How is ECF bleaching different from TCF bleaching?

    ECF bleaching still uses chemicals that contain atomic Cl, whereas TCF bleaching chemicals contain no atomic Cl whatsoever

16. True or False: Most bleached kraft pulp in the world is made using ECF bleaching.

    True

17. Explain how mechanical pulp bleaching is different from kraft or chemical pulp bleaching.

    Chemical/kraft pulp bleaching removes lignin, whereas mechanical pulp bleaching leaves most of the lignin in the pulp but decolorizes it

18. What is AOX and why is it important?

    AOX is adsorbable organic halide, and it tells how much chlorinated organic material is produced during bleaching; AOX is commonly limited by the EPA or another regulatory agency.

19. What is BOD and why is it important?

    BOD is biological oxygen demand, and it measures how much oxygen the effluent will use up in a river or lake if the effluent is not first treated; BOD is regulated by the EPA or other regulatory agencies)

20. What is meant by the "Cluster Rule" and when did it go into effect?

    The cluster rule is a U.S. EPA regulation for several things that are discharged from a mill; the rule went into effect in 1998

Chapter 8 - Paper Recycling

1. Define secondary fiber.

   Fiber stock that comes from a paper recycling operation.

2. Secondary fiber plants are usually located in areas of
   Why?

   Urban forests; facilities are close to collection and usage locations.

3. Name the basic grades of wastepaper.

   Mixed Paper; Old Newspapers (ONP); Old Corrugated Containers (OCC); Pulp Substitutes

4. What do the following acronyms stand for? a.ONP ; b.OCC

   (a) Old Newspapers (ONP); (b) Old Corrugated Containers (OCC)

5. Which type of wastepaper is used the most?

   Old Corrugated Containers (OCC)

6. Define pre-consumer wastepaper.

   Waste or scrap paper created in the manufacturing or converting process that has not been used by a consumer.

7. Define post-consumer wastepaper.

   Paper which has served its intended end-use and then discarded.

8. List the primary steps in handling wastepaper at a mill.

   Inspection, sorting, repulping.

9. List the six major process steps that may be used in contaminant removal.

   Separation (junk tower, ragger, detrashing). cleaning, screening, flotation, washing, bleaching

10. What are stickies?

    Contaminants such as hot melt glues, waxes, and tape adhesives.

11. List common contaminants in wastepaper repulping systems.

    Inks, stickies, dirt, (plastics, sand, metals, glues, adhesives, pigments, dyes, etc)

12. What is the purpose of deinking?

    To remove ink particles from the seconadary fibers

13. Ink particles are separated from ONP by either ______ or _______ steps or a combination of these.

    Washing or flotation

14. Name typical chemicals used in deinking.

    Detergents, dispersants, foaming agents and caustic soda (sodium hydroxide)

15. How does a flotation cell work for deinking?

    Air in the form of very small bubbles is mixed with the pulp slurry, and the ink particles attach themselves to the rising air bubbles. The foam that forms on the surface of the stock contains the ink particles, and this foam is skimmed off to separate the ink from the good fibers.

A

B

C

D

E